INSTITUTE OF ECONOMIC GROWTH

Studies in Economic Development and Planning No. 56

General Editor : K. Krishnamurty

STUDIES IN MINOR IRRIGATION
With Special Reference to Groundwater

Institute of Economic Growth, Delhi.

Studies in Economic Development and Planning

For a list of titles in this series published by the Institute of Economic Growth, see at the end of this book.

STUDIES IN MINOR IRRIGATION

With Special Reference to Groundwater

B.D. DHAWAN

1997
COMMONWEALTH PUBLISHERS
New Delhi

ISBN 81-7179-061-0

First published in 1990
Second, enlarged Edition 1997

Commonwealth Publishers
4831/24, Prahlad Street, Ansari Road,
Darya Ganj, New Delhi-110002 (India)

Laser Typesetted by :
Stylish Laser Typesetter, Delhi.

Printed at :
Efficient Offset Printers, New Delhi.

Dedicated To

The Unforgettable Memory of Ravi (1946-67)

Preface to the Second Edition

The subject of small-scale irrigation continues to be quite topical. I have been contemplating bringing out another edition of MINOR IRRIGATION book for quite some time. The first edition was brought out in 1990. As it was soon sold out, the publisher has been drawing my attention to the unmet demand. I am pleased to bring out this Second Edition.

The format of the book remains totally undisturbed. There is no point in revising a paper (which constitutes one chapter each in the book) that has been published earlier in a journal. Instead, I have added a new paper based on Census of Minor Irrigation (CMI). This paper was completed recently as an IEG study, entitled 'Minor Irrigation in India—A Critique' (February 1997). It is still unpublished, and is included here as chapter 13 with a changed title *viz.*, 'Census of Minor Irrigation: Some Results'. As usual, there was much delay in the release of *Report of Census of Minor Irrigation* (1994). Being an unpriced document of the Ministry of Water Resources, it has been little circulated. No wonder, it has not been systematically examined by the academic fraternity. Since the Report has no text which brings out highlights of the CMI, it is a tough nut to crack for those who are not well versed with minor irrigation subject. Admittedly, mine is a limited foray into its rich and varied contents.

Picture of minor irrigation segment of Indian irrigation sector is primarily constructed by the Indian planners on the basis of progress reports received from various governmental agencies associated with the development of such works. That ground realities could be at variance with such a picture was sought to be ascertained through a nationwide census of minor irrigation structures and their utilisation. Such a census, first of its kind, was carried out, with 1986-87 as the reference year. Some noteworthy findings are as follows:

1. Irrigation pumpset owners hardly got electricity for 4 hours in a day at peak time of crop seasons. This is mainly because

growth in electric irrigation pumpsets has outstripped the growth of electricity industry by a big margin. In the absence of requisite growth of electricity industry, the available supply of electric power is getting thinly distributed among pumpset owners.

2. Minor irrigation works other than tubewells turn out to be highly susceptible to drought. This is underlined by the poor performance of dugwells, tanks and other small-scale irrigation works in Gujarat during the census year which was a normal year for the country but was a severe drought year in Gujarat's case. This drought put almost half the minor irrigation works out of use (the corresponding national percentage was 13).

3. When we compare census numbers with corresponding official numbers, one discerns wide disagreement between the two for many States. To illustrate, the official number of electric pumpsets is one-third higher than the corresponding census figure. Again, in U.P. the official number of dugwells in existence turns out to be two and a half times the census figure.

4. Census figures of minor irrigation capacity and its utilisation turn out to be close to the official ones, but this is so at the national level, not at State level. Since the measurement of capacity is problematic, one should not read too much into the difference in two figures because both are measured in questionable manner. In fact, it is not very meaningful to seek such information from farmers through census. Instead, a specially designed sample survey would be more useful to assess unit irrigation capacity for various minor irrigation works. However, differences in actual area irrigated by minor irrigation works need to be viewed with right concern. Fortuitously, the direction of the difference is not the same everywhere. Plan figures exceed census figures by a large margin in U.P. (45 per cent), Orissa (122 per cent), Assam (141 per cent), but the difference is other way round in Haryana (-20 per cent), Maharashtra (-36 per cent) and Punjab (-31 per cent).

5. According to the census, about 1.6 million minor irrigation works were not in use. This suggests that plan estimate of created minor irrigation potential needs to be downgraded by a magnitude of 5.8 million hectares.

In bringing out this volume I am greatly aided by our Computer Unit, especially by Mr. Satya Narain who bore the brunt of word processing. For that, I am grateful to the Unit.

August 1997 **B.D. Dhawan**

Preface to the First Edition

Interest in groundwater irrigation heightened in the wake of HYV technology of the early sixties. Both in the popular media and within the academic portals agricultural growth came to be identified with groundwater development in general and private tubewell irrigation in particular. Lately, the growing opposition to the big dams has focussed further attention on minor irrigation, of which groundwater irrigation has been a leading component.

In a way, this volume, which is my third one on groundwater irrigation, might be viewed as a triology on groundwater economics in India. It is a collection of papers published in different journals. I am sure bringing them together in one place would be of particular interest to research scholars who find locating journal articles a bit time consuming and tedious task. In view of this referential angle, the papers are reproduced without any modification.

Aside from the introductory chapter one, the volume consists of eleven chapters, each constituting a paper in itself. While chapters 2 and 3 deal with groundwater endowment and its irrigation potential, the management aspects of this precious resource are taken up in chapter 4. A summary view of the economics of tubewell technology, the most preferred among the groundwater technologies, is given in chapter 5. The evidence on scale economies of this technology is analysed in chapter 6. A comparative picture of traditional and modern groundwater technologies, alongwith their external effects, is dealt with in chapters 7 to 9. The problems of groundwater development in Uttar Pradesh, the largest state of the Indian Union, are discussed in chapters 10 to 12. The papers in these chapters are based on two fairly large studies, namely, *Groundwater Irrigation in East U.P.* (IEG, mimeo, 1978) and *Underutilisation of Minor Irrigation Works : A Case Study of State-owned Tubewells* (IEG, mimeo, 1969), prepared by the author at the Institute of Economic Growth.

Admittedly, the key ideas of the papers in this collection have filtered into my four books on irrigation published under the IEG auspices during the eighties. In point of fact, these papers are duly listed in the bibliography section of the aforesaid books. But the aforesaid books are written in a format quite different from the one used in research papers, wherein a good deal of attention is paid to data sources, footnoting and other technicalities that are of considerable concern to fellow professional research scholars. By faithfully reproducing the original papers in a compact volume the author hopes to add to the empirical literature on groundwater that can be readily tapped when in book form.

In the preparation of this volume, I have been greatly assisted by my friend B.D. Kapoor and by Mrs. Rama Sethi. Though I cannot enumerate all the research assistants who competently and ungrudgingly helped me write the original papers over the last two decades, I am not oblivious of their role in applied economic research.

B.D. Dhawan

Contents

Contents

List of Tables

1

Introduction

Irrigation occupies a signal position in Indian agriculture. In its absence Indian farmer has revealed a marked disinclination to make intensive use of the scarce land resource. Thus land use during the dry period of the agricultural year gets extremely restricted without irrigation. And the fear of moisture stress due to a break in the monsoon inhibits him from the use of chemical fertilisers, the leading input of modern agriculture.

Irrigation works in India are classified in three categories : (1). minor irrigation, (2). medium irrigation, and (3) major irrigation. This three-way classification was evolved during the British times, and was linked to the quantum of capital resources required for executing irrigation works of varying scale or size. When the resource requirement for an irrigation scheme was sizeable, it was categorised as a 'major irrigation work'. Likewise, if this requirement was of a small order, the scheme in question became a 'minor irrigation work' in official parlance and in public accounts of the government – in-between came the medium irrigation category. The cut-off points for the investment norm were occasionally revised following price inflation. For example, this norm for minor irrigation schemes was raised from Rs. 15 lakhs to Rs. 25 lakhs in 1970 (Rs. 30 lakhs for hilly areas). Finally, this financial norm was abandoned altogether by the Indian Planning Commission in early 1978.

Now the categorisation of irrigation schemes is on the basis of their culturable command area (CCA) as per the following norms :

minor work	less than 2000 ha of CCA
medium work	2,000 – 10,000 ha of CCA
major work	10,000 ha and above of CCA

While all surface water-based schemes are classified by the above-mentioned categorisation, every groundwater-based irrigation scheme, whether of the public or of the private sector, is deemed to be in the minor irrigation segment. More specifically, minor irrigation works in Indian planning comprise the following fifteen types of schemes :

I. Groundwater Related

1. new open or dugwells
2. new tubewells
3. filter points
4. artesian wells
5. water-lifting appliances
6. well boring
7. major repairs to wells
8. deepening of existing wells

II. Surface Water Related

9. tanks
10. diversion schemes including 'kuhls' of hill areas
11. lift irrigation
12. flood protection works
13. drainage works
14. anti-sea water intrusion works
15. water conservation-cum-groundwater recharge works.

Minor irrigation works, taken in totality, have been of great importance in Indian irrigated agriculture. Their irrigation potential today is estimated to be of the order of 46 million hectares (of crop area). This considerably exceeds the corresponding estimated potential of nearly 33 million ha from major and medium irrigation works. However, it is not correct to assert that this excess is of recent origin. As far back as the commencement of the First Five Year Plan in 1951, the minor irrigation works accounted for 12.9 million ha out of total national irrigation capacity of 22.6 million ha. Therefore., the coinage of the catchy phrase 'minor becomes major' by B.B. Vohra in the early seventies and its vigorous propagation in recent years in support of minor irrigation cause – should not be deemed as a new landmark in the development of Indian irrigation.

During the course of nearly four decades of economic planning in India, most of the additions to minor irrigation capacity have been from groundwater sources. More specifically, while the created minor irrigation potential from surface water sources is officially estimated to have risen from 6.4 to 11.1 million ha in the aforesaid period, the same potential from wells zoomed from 6.5 to 34.8 million ha in the same period. It is because of such a predominant place of groundwater irrigation within the minor irrigation segment that the former has come to be identified with the latter in the minds of Indian water planners.

Some Simplistic Impressions

As a society switches from traditional to modern ways of doing things in the production as well as the consumption realm, a lingering doubt or suspicion arises about the new in some minds and a sort of romanticism grows around the old. Till the rise of pumpset technology for lifting groundwater the two main modes of minor irrigation, namely, tanks and dugwells, epitomised traditional irrigation in comparison to modern canals which originate from large engineering feats either in the shape of big barrages on rivers or from huge storage reservoirs created by damming big rivers at gorge points in a river basin. No wonder, then, that some halo has come to surround minor irrigation. In what follows I shall briefly comment on three impressions of questionable veracity.

To begin with, there is the impression that the Planning Commission has neglected the development of minor irrigation works. For, it has allocated far more resources to major and medium irrigation schemes – one estimate places the comparative picture of resource allocation in the ratio of 5 to 1. This impression has arisen because of a serious misreading of our financial data as published in either budget papers of the state and central governments or in Five Year Plan reports of the Planning Commission.

Budgetary outlays on investment schemes pertain to public (i.e. government) sector only. As many of the development programmes for minor irrigation, unlike in respect of major and medium irrigation, are in the private sector, their corresponding investment outlays on farmers' account are not indicated in the budget documents. Even reference to their investment requirements in the plan documents is *partial*, in the sense that only public and institutional sources of finance for them are spelt out in a given plan. Because of the exclusion of private investment figures, the budget documents grossly understate the scale of minor irrigation planning activity vis-a-vis major irrigation – the degree of understatement is much lower when the Five Year Plan documents are used to gauge the investment outlays. It is only when one takes recourse to the data given in the background reports of the working groups on minor irrigation for each plan period that one gets a complete picture of the investment outlays (inclusive of private outlays) on minor irrigation in Indian irrigation planning. To illustrate, the total investment requirements for minor irrigation for the Eighth Plan are reckoned at Rs. 141 billion (Rs. 14,145 crores) as compared to Rs. 264 billion (Rs. 26,430 crores) for major and medium irrigation.

Another facile impression, which is related to the one mentioned already, is that the resource cost of developing minor irrigation is low. Evidently, if partial figures of investment outlays given in either budget or plan documents are divided by the statistics of irrigation potential for the minor irrigation segment as a whole (i.e., irrigation works of both the public and private sectors), one would end up with underestimation of unit investment cost per hectare of irrigated area potential for minor irrigation works. Even if one takes recourse to correct and complete data on investment, one is not reckoning with the huge supporting investments of groundwater irrigation. Among these supporting investments the most important ones are those on account of power generation and its transmission to pumpsets in distant villages.

Electric pumpset has been the most popular waterlifting device. Their numbers have risen phenomenally since 1951. They numbered 5.7 million in the mid-eighties, and are likely to have crossed the 8 million mark by the end of the eighties (i.e., the end of the Seventh Five Year Plan). The investment requirements for power generation and its supply to pumpsets in rural areas is very substantial – and probably more than the direct outlays incurred on minor irrigation works equipped with electric pumpsets. Without duly reckoning with such supporting investments it is unwarranted to infer that minor irrigation development entails lower capital costs than does major irrigation.

Finally, there is considerable misconception about capacity utilisation in respect of minor irrigation works. Lionisation of this mode of irrigation by the anti-big dam lobbyists is misplaced to the extent it is perceived to be free from the problem of underutilisation of created capacity. The illusion of full capacity utilisation of minor irrigation potential has probably arisen because of a *working assumption* of our planners in this regard. Now that this assumption has been given up we no more see a picture of full capacity utilisation for minor irrigation works in the plan documents. For instance, about 2.2 million ha of minor irrigation capacity was unutilised in the mid-eighties. Since then, about one million ha out of about 8.4 million ha of new capacity addition made during the Seventh Plan period is feared to have remained unutilised.

Yardsticks of Progress

As irrigation is created with the twin-purpose of augmenting crop production and protecting farm economy from the periodic drought, the ideal summary measure of the progress in minor irrigation development would probably be a time series of aggregate crop production/farm income originating from lands watered by all minor irrigation sources taken together. The reality being far from the ideal situations, one has invariably to judge progress through a variety of measures bearing some affinity with minor irrigation : cumulative number of tanks, tubewells, dugwells, pumpsets etc.; area irrigated by tanks, tubewells and dugwells; investment in and institutional finance/ government subsidies for minor irrigation works; growth in the number of pumpsets/villages electrified and electricity consumption in the agriculture sector; so on and so forth. For the reasons given below, measuring minor irrigation progress through the mirror image of financial statistics needs to be eschewed unless one can take pains to grapple with the drawbacks of this otherwise good summary measure of practical significance in the planning work.

The author has already pointed out (a) the partial character of investment data on minor irrigation given in budget as well as plan documents and (b) the pervasive phenomenon of not reckoning with substantial supporting outlays not figuring in the minor irrigation head of such documents. Regarding the attempt by the Working Groups on Minor Irrigation to estimate the investment on private account, his apprehension is that such estimates are very much on the low side. Due care is not taken in these Groups (dominated by the Chief Engineer, Minor Irrigation, GOI) to assess such investment correctly simply because the resultant figures are of no practical significance in plan resource allocation exercises.

In fact, too much reliance is placed, while making such estimates, on the margin money norms adopted by Indian financial institutions for financing/refinancing investments in minor irrigation schemes. This leads to underestimation on two counts. For one, the 15-20 per cent margin of funds from farmers' own funds (including his borrowings from friends, relatives and money lenders) is of little relevance to new wells constructed without financial/government funds – half the wells/pumpsets, according to one informed source, are financed in this manner. For another, as the actual construction costs of wells tend to exceed the investment norms used by institutions, the farmers' own financing overshoots their margin amount – the relative proportion of farmers' own funds *vis-a-vis* institutional funds in a well can rise substantially if the eligibility criteria does not permit requisite loaning to a farmer (occasionally this criteria is wrongly enforced at the primary lending agency level). Thus, the needed investment on private account for the 8th Plan is placed at Rs. 1440 crores, which is a little over one-tenth of the total financial outlay for the whole minor irrigation segment.

Because of persistent price inflation the rising outlays on minor irrigation exaggerate the progress on minor irrigation front. Therefore, such nominal figures need to be brought first in constant prices, so as to appropriately reflect the growth in minor irrigation. The use of the general price deflator may be less satisfactory for nominal investments in pumpsets and tubewells whose prices, capacity for capacity, have lagged behind the general price inflation. However, the more serious problems here are on two following counts.

One is the absence of any information on capital stock for the pre-planning point viz. 1951. The other is due to the fact that a portion of fresh investments is in the nature of either a replacement for an earlier asset or a total displacement of an earlier mode of minor irrigation by a more modern mode e.g., the abandonment of dugwell irrigation in the Indo-Gangetic plains by farmers who installed their own tubewells.

At least the first problem of initial capital stock does not arise in the case of the other summary measure, namely, progress in minor irrigation potential in terms of crop area that can be potentially irrigated. Therefore, area potential statistics reflect progress in minor irrigation more correctly than do investment statistics. However, this yardstick too has an element of overstating progress. This may be attributed *inter alia* to certain inherent drawbacks in the methodology of cumulation used in arriving at the estimates of minor irrigation potential (or the numbers for various kinds of groundwater structures). Such drawbacks are briefly outlined below, alongwith the direction of implications.

Cumulation is simply additive arithmetic, whereby irrigation potential purportedly created during a given period is superimposed on that already arrived at, also through cumulative process, for the beginning of that period. Obviously, such cumulation abstracts from the following three negative features that do detract it, making the net addition to capacity less than the gross addition :

(i) An old minor irrigation work falling into disuse much before its useful life is over;
(ii) A new minor irrigation work becoming simply a supplementary(or standby) source of watering to another minor/major source of irrigation without commensurate expansion in irrigated acreage;
(iii) Reduction in irrigation capability of minor irrigation works due to aging process plus poor repair/preventive maintenance work.

Admittedly, it is difficult to quantify effects of such temporal developments. Still, two observations may be in order. First, while the aging process erodes irrigation capability of a work very gradually, its impact can remain imperceptible for long only if preventive maintenance and timely repairs are effected. But if repair and maintenance activity is woefully below par – as is the case with government - maintained tanks or government-owned tubewells – the capacity erosion occurs faster than one due to normal wear and tear. In the case of private tubewells one may largely condone the non-allowance for depreciation in rated capacity. The saving feature here, as per this author's apprehension, is that the adopted irrigation potential norm per private tubewell is reckoned well below its intrinsic or technical capacity limit which can be realised in the absence of constraints on the supply and demand side for tubewell water.

Second, the non-additivity of irrigation capacity of major and minor irrigation works has been known since long. A good deal of the canal irrigation development that took place during the British rule in the well-populated tracts of the Indian sub-continent simply displaced traditional groundwater and surface water irrigation sources already existing there. But the phenomenon of conjunctive use of our water resources is of more recent origin. Farmers within canal and tank commands now increasingly invest in well irrigation. Such investments also serve the purpose of supplementary/standby source of watering crops during the main season when main canals are running and tanks are not yet fully depleted. Such conjunctive use adds much more to overall crop production rather than to total irrigated area. This problem of non-additivity of irrigation capacities exists both between major and minor irrigation on the one hand and among minor irrigation sources themselves on the other.

Bearing in mind the cautionary observations made in the foregoing paras, let us briefly look at some salient trends and tendencies in Indian minor irrigation. The main yardsticks of progress in minor irrigation are listed in Tables 1.1 and 1.2. While the data for the Seventh Plan is subject to revision at a later date when actual figures for the last two years of this Plan become available, the Eighth Plan data are simply the proposals of the Minor Irrigation Group for this Plan which submitted its report in July 1989 – such proposals are normally reviewed downwards by the Planning Commission.

TABLE 1.1

Progressive Stock of Groundwater Structures in India

(in thousands)

Year	*Dugwells*	*Shallow tubewells*	*Public tubewells*	*Electic pumpsets*	*Diesel pumpsets*
1950-51	3,860	3	2.4	21	66
1960-61	4,540	22	8.9	200	230
1968-69	6,100	360	14.7	1,090	720
1973-74	6,700	1,138	22.0	2,430	1,750
1977-78	7,435	1,740	30.0	3,300	2,350
1979-80	7,786	2,132	33.3	3,965	2,650
1984-85	8,742	3,359	46.2	5,709	3,550
1989-90*	9,487	4,754	63.4	8,226	4,355
1990-95**	11,198	6,443	74.4	11,226	4,977

* anticipated

** based on proposals

In terms of numbers, the single most important minor irrigation work is still a dugwell in spite of its comparative slower expansion during the period of planning. Dugwells are now modern in character, in the sense that these are increasingly equipped with pumpsets and have pit borings at the bottom for enhancing well water yield. Next in importance are shallow tubewells whose numbers have literally exploded since the mid-sixties. These now number over 4.5 million (as compared to nearly 9.5 million dugwells), and are wholly in the private sector. There are also large capacity tubewells, numbering over 60,000, in the public sector. Bulk of the tubewells are located in the Indo-Gangetic plains, with Uttar Pradesh alone accounting for nearly half of them.

The stock of electric pumpsets is now almost twice as much as the diesel pumpsets. Some of these pumpsets are installed on surface water sources. There were, according to preliminary tabulations of the Census of Minor Irrigation Works, about 0.17 million surface water lift irrigation schemes during 1986-87. Tanks numbered around 0.16 million, surface water diversion schemes (using the force of earth's

TABLE 1.2
Development of Minor Irrigation in India

S. No.	*Period*	*Created irrigation potential (million hactares)*			*Investment in crores of rupees*		
		Surface water	*Ground water*	*Total*	*Plan outlay*	*Credit*	*Total*
1	*2*	*3*	*4*	*5*	*6*	*7*	*8*
1.	Ultimate	15	40	55			
2.	Pre-Plan	6.4	6.5	12.9			
3.	First Plan 51-56	6.43	7.63	14.06	66	Neg.	66
4.	Second Plan 56-61	6.45	8.3	14.75	142	19	161
5.	Third Plan 61-66	6.48	10.52	17.00	328	115	443
6.	Annual Plan 66-69	6.50	12.50	19.00	326	235	561
7.	Fourth Plan 69-74	7.0	16.5	23.5	513	661	1174
8.	Fifth Plan 74-78	7.5	19.8	27.3	631	780	1141
9.	Annual Plan 78-79	7.75	20.85	28.6	237	240	477
10.	Annual Plan 79-80	8.0	22.0	30.0	260	250	510
11.	Sixth Plan 80-85	9.7	27.8	37.5	1802	1438	3240
12.	1985-86	9.99	29.03	39.02	493	501	994
13.	1986-87	10.27	30.38	40.65	588	584	1172
14.	1987-88	10.54	31.13	42.27	642	610	1252
15.	1988-89*	10.85	33.13	43.99	739	773	1512
16.	1989 -90*	11.15	34.78	45.94	766	843	1609
17.	Eighth Plan 90-95**	13.26	44.79	58.06	7513	5119	12632

* anticipated

** proposed.

gravity) totalled 0.09 million. Together, the combined irrigation potential from these three types of surface water-based minor works now stands around 11 million hectare which is not even twice that of its pre-planning level of 6.4 million ha. In comparison, the irrigation potential from groundwater sources has multiplied five-fold, from 6.5 million ha in 1951 to over 34 million ha in 1990. Institutional funds have played a signal role in the expansion of all minor irrigation works – this source of financing is now of the order of Rs. 800 crores per annum.

To conclude, one does not discern any tell-tale marks of any decelerating progress in minor irrigation during the eighties vis-a-vis the seventies. That there is no deceleration in absolute terms in minor irrigation expansion is borne by the following incremental additions per annum :

	Minor irrigation potential created (000 ha)	*No. of groundwater sources (in 000)*			
		shallow tubewells	*dugwells*	*electric pumps*	*diesel pumps*
First Plan	232	} 2	68	18	16
Second Plan	138				
Third Plan	450	16	112	63	48
Annual Plans (66-69)	666	86	na	192	83
Fourth Plan	900	156	120	268	206
Fifth Plan	950	151	184	218	150
Annual Plan (1978-79)	1,300	200	180	300	150
Annual Plan (1979-80)	1,400	192	171	365	150
Sixth Plan	1,500	245	191	349	180
Seventh Plan	1,688	279	149	503	161
Eighth Plan (proposals)	2,424	338	342	600	124

Proposals for Eighth Plan

The programme for minor irrigation development for the Eighth Plan period, as per the recommendations of the Working Group on Minor Irrigation, is bolder than the one implemented during the last Plan. The extra step-up in the physical targets may be justified on a perception that, owing to the increasing opposition to the big dam idea and the acute paucity of investible funds for public sector schemes, there may be serious shortfalls in major and medium irrigation. However, the feasibility of accelerated development needs a closer look.

By feasibility I do not mean simply the availability of both financial and other resources required for the construction of minor irrigation works. The financial requirements of the recommended programme do reveal a very substantial step-up over the Plan gone by. More specifically, the estimated outlay requirements for Eighth vis-a-vis the Seventh Plan are as follows:

	Seventh Plan	Eighth Plan
	(Rs. in crores)	
1. Public sector :		
(i) States	3,076	6,015
(ii) Centre	151	1,498
total (i+ii)	3,227	7,513
2. Institutional finance	3,312	5,119
3. Private sector	1,980*	1,400

* As visualised by the Seventh Plan Working Group.

Because of the new regime's commitment to allocate half the plan funds to rural economy, one can hope that the requisite public sector outlays would be forthcoming for minor irrigation works. It is the new regime's election manifesto of providing substantial debt relief to farmers (by waiving their loans to the extent of Rs. 10,000 crores) that can jeopardise institutional financing for minor irrigation schemes during the coming years. In case apprehensions about credit availability do materialise, farmers would have to put in more of their own resources for establishing well irrigation under their own aegis. The prospects for such extra mobilisation are not dim because the farmers have now been promised much better deal in farm product prices.

My anxiety with regard to feasibility of the proposals is on the following three counts. First, the pace of minor irrigation development is vitally linked with the pace of rural electrification and power generation programmes, both of which are highly capital using infrastructural activities. There were sizable shortfalls in pump energisation target during the last Plan. Despite impressive progress in power generation, power shortages, both in quantum and quality, plagued the countryside – in Uttar Pradesh the World Bank - financed public tubewells, which are supposed to get 16 hours of uninterrupted daily power supply through special power lines, hardly get power for 8 to 10 hours (6 hours according to some sources) per day, thereby crippling their working.

Second, the regional break-up of the recommended proposals is not in harmony with our federal polity. Uttar Pradesh has been singled out for a questionable share in the total programme. This state is to account for 5.7 million ha out of the total 10 million ha of additional potential from groundwater sources. If availability of groundwater resource is a relevant consideration, then why allow only 0.4 million ha capacity expansion for West Bengal? Why visualise 10 lakh out of 17 lakh new shallow tubewells in U.P., and only one lakh for West Bengal ?

Third, the untapped groundwater resource of the nation poses difficulties in exploitation under private aegis. How small size of land holdings and their fragmentation in space militate against the development of individually-owned tubewells in the groundwater–abundant east Gangetic plains is well known. In the hard rock regions, the unexploited groundwater potential is firstly not that large as is implied by the estimates in this regard, and secondly it is primarily located in tracts served by canals and tanks. Such seeped-in waters from surface irrigation works have not been tapped in the upper reaches of canals wherein farmers have managed to meet their irrigation needs primarily from canal supplies. Such farmers have little inducement to invest in their own wells. Such inducement can, however, be created by ensuring appropriate management of canal system, whereby the farmers are prevented from drawing on extra canal supplies at the expense of their brethren in the tail-end reaches - the latter have readily invested in well irrigation because of shortage of canal supplies.

About the Papers

Readers must bear in mind the time element which is of some contextual significance in appreciating the research contribution of a paper. These papers have a strong empirical content which make them a rather heavy reading material – the overlap in some papers might lighten the burden of reading. Paper by paper, I have the following sum-up to offer for sustaining readers' interest.

The *ad hoc* character of the estimates of groundwater resources even as late as 1976 is highlighted in chapter 2. Besides drawing attention to the nature of infirmities in such estimates and the official attempts underway then to remedy the same, two central points have been stressed in this chapter. First is the need for keeping distinction between stock and flow dimensions of groundwater resource. From the stock viewpoint, groundwater is an immense resource. Prudence in groundwater planning demands that its utilisation should be limited to its annual replenishment, technically known as groundwater recharge. Utilisation in excess of this long-run rate is permissible where water table is high (i.e. waterlogging conditions exist), or when calamity of a drought occurs. But the excess withdrawals in the drought year need to be compensated afterwards by underdrawal in normal and above-normal rainfall years. Second, a substantial component of groundwater recharge is from surface irrigation works. As the nation develops more of the irrigation potential from its river/stream flows, the volume of groundwater potential available for exploitation rises. Thus, groundwater resource is dynamic, not static. That is to say, our irrigation potential from groundwater is not fixed by nature but grows in volume due to man-made efforts in the shape of incidental recharge from canals and irrigation tanks – also through artificial recharge from percolation tanks, soil and water conservation works etc.

In chapter 3 we critically look at the new estimates of irrigation potential from our groundwater resources. The revision has been done very recently by the respective state groundwater agencies, which were set up in the seventies and which have done useful work in monitoring fluctuations in groundwater table. The apex groundwater body, namely, the Central Groundwater Board, has played a key part in the revision. However, the revised estimates are still far from firm. The apprehension is that these are on the high side. The hazards of groundwater development on the basis of rosy estimates of groundwater potential need to be borne in mind. The reassessment of our groundwater irrigation potential must take place without any extraneous motives, such as, diverting planning away from surface water works to groundwater-based irrigation, or ensuring the flow of institutional and World Bank funds for groundwater development. The need for a closer scrutiny of the new estimates is emphasised more so in the case of states underlain with hard rocks or having low rainfall where groundwater resource is naturally a scarce and fragile resource.

Excessive exploitation of groundwater is a matter of wider concern, transcending environmentalists as well as hydrogeologists – the latter have been pleading for groundwater regulation for two decades. Conservation of this 'common property resource' is the watchword of the day. Fervent pleas from diverse quarters are made for legislative enactments in this regard. The author argues in chapter 4 that such enactments, even if these witness a passage through our state legislatures, are of little practical relevance in the Indian conditions. Instead, considerable milage can be achieved through indirect measures like appropriate pricing of electricity for rural pumps. The set of such indirect measures need to be broadened.

The short paper in chapter 5 offers an insightful view of the economics of tubewell technology. This technology for tapping groundwater was readily adopted by the Indian farmer as its profitability rose manifold following the advent of high yielding variety (HYV) seeds of wheat and paddy. Groundwater extraction capacity of a tubewell being much more than that of a traiditional well, excessive reliance on tubewell irrigation puts a built-in pressure on groundwater resources. But for extensive development of canal irrigation in the western Indo-Gangetic plains the explosive growth of tubewell irrigation would have put the groundwater resource under siege here.

The issue of scale economies in tubewell irrigation, which is briefly dealt with in the preceding chapter 5, is investigated in detail in chapter 6. The study of this issue is certainly not of much relevance from the viewpoint of private tubewell irrigation as Indian farmers, unlike in the western U.S.A., possess rather tiny and fragmented land holdings. Its practical importance stems from the viewpoint of setting up large capacity tubewells in public domain. One does not find a U-shaped unit cost curve; it is more like an L-shaped curve. The optimum size for a public tubewell is governed not by pure internal scale economies but also by considerations of its external effects on farmers using their own means of groundwater irrigation.

The hallmark of modern times is the rapidity of change, the modern displacing the traditional, the subject of enquiry in chapter 7. Modernity among the various modes of well irrigation may be graded in the following descending order, in which an electric tubewell is rated the most modern while a traditional well operated with a human-powered mechanism as the most traditional :

1. electric tubewell
2. diesel tubewell
3. electric dug-cum-bore well
4. diesel dug-cum-bore well
5. dugwell equipped with an electric pumpset
6. dugwell equipped with a diesel pumpset
7. dugwell equipped with an animal-drawn persian wheel
8. dugwell equipped with a 'mhot' (animal-operated)
9. dugwell operated with human-powered waterlift.

The main objective of this paper is to reappraise, both from private and social angles, the modern groundwater irrigation vis-a-vis the traditional one. The paper is divided into two sections. The techno-economic features of the two groundwater technologies, discussed in section I, provides the necessary backdrop to the discussions in the section II on external diseconomies imposed by the new on the old technology.

The internal and external economies of groundwater exploitation in India are investigated in considerable empirical detail in chapter 8. How the new pumpset technology affects adversely the welfare of small farmers using traditional modes of well irrigation is taken up in chapter 9. The special problems of groundwater exploitation in East Uttar Pradesh, an important plains region of the vast Gangetic basin, are analysed in chapter 10. In its Section I the paper reviews trends in groundwater irrigation, whereas small and large farmers' preferences among various types of wells are analysed in section II.

The last section III considers the wide array of factors that promote or impede groundwater development.

One major drawback in water pricing in India that has been pointed out by innumerable economists is that farmers are not charged for canal supplies on volumetric basis, thereby providing them with no inducement to use water efficiently/economically. In the case of farmers getting water from public tubewells it is possible to charge them on proximate volumetric basis. The paper in chapter 11 examines how far the demand for tubewell water is influenced negatively by the price of water, or positively by the price of farm product produced with that tubewell water. Econometric evidence reveals that farm product price is far more dominant determinant of farmers' demand for irrigation water than is the pitch of the water tariff.

The concluding paper in this volume is about the problem of unutilised capacity existing in public tubewells. Hardly two-fifths of the irrigation potential of these tubewells is presently utilised -- this fraction is not even one-tenth for the state of Bihar. This gross underutilisation, which is far more serious than is the case in respect of public canals or major irrigation works, speaks volumes for the poor management of public tubewell irrigation. The phenomenon of underutilisation is indeed complex. It is widely felt among the well-informed circles that removal of the crippling constraint of power availability alone would substantially eliminate the problem of unutilised capacity in state tubewells. To a large extent, the mismanagement in the electricity supply system, which also is in the public domain, is making management of public tubewells difficult. If a single factor is to be identified for the malaise, it is the callous disinterestedness of the state leadership, both of the political and the bureaucratic realms that militates against the full realisation of the intrinsic economics of public tubewells in the Indian conditions of small and fragmented land holdings.

2

India's Groundwater Resources*

Canal irrigation has played a vital role in transforming the precarious agriculture of semi-arid regions – particularly in north-west India -- into productive agriculture over the last 200 years. Now it is groundwater which is playing a critical role in our agricultural transformation, ever since the advent of high-yielding variety (HYV) seeds. In fact, innumerable research investigations have highlighted a close relationship between the success of HYV programme in an area and the use of groundwater irrigation -- especially individually-owned tubewells and wells fitted with power pumps.

It is known that well irrigation was abandoned by many farmers when canal irrigation was introduced in their area. Two main factors responsible for this behaviour of the irrigators were : (1) the comparatively low cost of canal irrigation *vis-a-vis* traditional well irrigation, and (2) the not so exacting water requirements of traditional or desi crop varieties *vis-a-vis* HYV seeds. The traditional varieties had evolved through a process of natural selection, during cultivation extending over centuries, and had thus acquired capacity to withstand substantial variations in water application. Whatever dent was made in their crop yield by untimely water supplies from public canals was probably more than offset by the sharp economy in irrigation cost following the switchover from well irrigation to canal irrigation. The differential in respect of irrigation cost would be particularly large for Rabi and Zaid (summer) seasons, when the water table is lowerer than that during Kharif season. For, the marginal cost of lifting water with manual or animal-drawn lifts rises sharply in an open well with an increase in the depth of water table.

---

* Reprinted from *Economic and Political Weekly,* Vol XII No. 13, Review of Agriculture, March 1977

The advent of HYV seeds has once again revived interest in owning one's own source of irrigation so as to meet timely irrigation needs of the new varieties. Since these varieties are not a product of natural selection, they are bound to lack in hardiness for coping with vicissitudes of weather or with untimeliness in agricultural operations. In view of the critical importance of water in their agronomical regime they have imparted a tremendous spurt to investment in groundwater irrigation.

For a variety of reasons – such as (a) concentration of agricultural development effort in areas already irrigated and (b) better availability of farmer's own investible funds in canal-irrigated areas -- one now finds that farmers in canal-irrigated tracts have evinced very keen interest in developing their own sources of groundwater irrigation. This conjunctive use of groundwater with surface water in canal tracts is a very welcome development from the viewpoint of optimal use of our water resources. The continued use of groundwater sources is the real answer to the long-run danger of water-logging and consequent soil salinity in canal-irrigated tracts. Where farmers are too poor or too small to undertake investment in groundwater irrigation, the State can invest in large tubewells which would also augment and firm up canal water supplies. Such augmentation tubewells are now being planned in Uttar Pradesh.

It can be easily demonstrated that sources of public irrigation are inherently incapable of assuring timely irrigation to every farmer in their commands. Even if their administrators are equipped with necessary data about each farmer's farm and are provided with computer facilities to handle the mass of farm information about their irrigation commands, they would still be unable to devise water delivery schedules so as to meet the irrigation needs of *each crop of every farmer*. It is true that this intrinsic problem of lack of timely irrigation can be minimised considerably if (a) the size of the public irrigation project is kept small, (b) its culturable command is deliberately reduced so as to provide intensive or productive irrigation instead of protective or thin irrigation per unit of irrigable land, and (c) the project is strengthened with a proper storage reservoir (not needed in case of public tubewells) that would cope with the vagaries of the weather – the ultimate source of practically every macro dis turbance in the economy during peace time. But, then, these conditions are difficult to fulfill if technical and egalitarian considerations loom large while making the decision on a project.'

The HYV seeds, unlike the earlier or *desi* seeds, perform well only when pampered well with requisite inputs and care. In view of their exacting demand for water, ensuring timely irrigation for them is almost impossible unless a farmer has control on the source of irrigation. In order to attain complete control on a water source its ownership is a must, a condition that is easily fulfilled by groundwater as compared to surface water (ponds and small tanks being minor exceptions). However, ownership of source is only a necessary, but not a sufficient, condition for ensuring timely irrigation. In the event of failure of rains, dug-wells become an unreliable source of irrigation – i.e., precisely when there is pressing need for irrigation. Likewise, tubewells, which are relatively less susceptible to drought, can become an unreliable source of irrigation if the electricity supply system is inefficient and inadequate. In such a situation, diesel tubewells appear an attractive proposition for supporting HYV cultivation -- though these tubewells, too, make rather exacting demands of repair and preventive maintenance services not readily available to distant villages. In short, ownership of a groundwater source is no guarantee for timely irrigation to a farmer operating in an environment of underdeveloped infrastructure – comprising electricity, roads, repair and maintenance skills, etc.

Whereas the advent of HYV seeds has generated widespread appreciation of groundwater resources, some fanciful and some totally erroneous notions do persist about this vital resource. For example very many people are under the illusion that unlimited water supplies exist below the land surface : if wells cannot reach this water, dug-cum-bore wells or tubewells should be constructed. They do not know that, quite often, deeper strata of groundwater are saline and brackish wherever scanty water is encountered in the upper portions of the ground. Such saline water, if used for irrigation, can lead to salt accumulation on soil surface, thereby rendering land unfit for cultivation before long.

In appraising groundwater resources, it is imperative to keep a clear distinction between *stock* and *flow* dimension. From the stock viewpoint, groundwater is probably an immense resource in many parts of India. But, judged from the flow angle, it is a bountiful resource in the eastern Gangetic plains only. When some people quote approvingly but out of context an expert who might have said that the volume of groundwater below the land surface of Uttar Pradesh is much more than the volume of water flowing annually in the Ganges and its tributaries, or that the water underneath UP is equivalent to hundreds of Bhakra reservoirs, they are taking a stock or capital view of the groundwater resource. But when some concerned hydrogeologists caution against overexploitation of groundwater resource, they are taking a flow or 'safe yield' view of groundwater resource. The two views can generate quite contrary impressions or reactions, unless one is on the guard as to whether the stock or the flow view is implied. Let me illustrate the point with the help of a concrete example.

UP plains are said to be underlain with alluvial deposits upto a depth of 5 to 6 thousand metres at some places. Since it is these deposits of sand and clay which really yield groundwater, one may say that, in UP groundwater occurs upto a depth of 20,000 feet. However the volume of this underground water, as per cent of the volume of alluvial deposits, would be no more than 10 per cent – 15 to 20 per cent in the case of sandy aquifers and 3 to 5 per cent in the case of clayey aquifers. In terms of volume, the stock of groundwater may be as much as 2,000 million hectare-metres (mhm) if we restrict ourselves to the top 1000 metres of alluvial deposits. Even this stock of groundwater under the UP plains comes to a figure which is over 30 times the estimated annual flow of 55 mhm of the entire Ganges river system. Compared to this massive figure, the groundwater (flow) potential of Uttar Pradesh is reckoned at no more than 14 mhm by any expert. For, this potential is worked out on the 'safe yield' principle – i.e., determining the volume of water which annually flows into the ground (below crop root zone, say 10 feet below the land surface) and which can be safely extracted without any untoward decline in water table.

The stock of groundwater has been built gradually from times immemorial. Evidently, this water cannot be exploited on the same principles as other mineral resources, for which new sources of deposits or substitute minerals are often discovered before long. Prudence requires its utilisation at a long-run annual rate of no more than the annual recharge of the groundwater stock. Exceptions from this prudent policy are conceivable in two circumstances.

One such circumstance is when there is water-logging in the crop root zone, which tells upon crop production. Despite the low intensity of irrigation in canal commands, considerable water-logging has occurred in the old canal areas. Overdraft above annual recharge would be a sensible policy in such areas so long as the water table is lowered appreciably below the crop root zone. The other permitting circumstance for overdraft on groundwater stock may be one in which prolonged overdraft gives rise to a new equilibrium level of water table. This once-for-all depletion in groundwater resource may be beneficial in areas which are criss-crossed by rivers and streams underlain with coarse sand deposit, with the result that 'induced' recharge of groundwater goes up sufficiently so as to match the overdraft. In a way, this is essentially a scheme for augmenting the groundwater potential. However, it needs to be implemented with great care. One has to guard against lowering of the water table below the bed level of a river; otherwise the 'base' (non-monsoonal) flow of rivers may disappear altogether. This can create serious problems for those who benefit from the off-season flow of the rivers. In particular, the fish population of the rivers can be seriously threatened and the disposal of city and industrial liquid wastes can become a big problem. Also, the water table should not be stabilised at such a great depth that either it becomes uneconomic to lift water with private means from such a depth, or saline water from the neighbourhood infiltrates the emptied space. This last danger exists in areas lying close to the sea or to the deserts which are underlain with saline waters.

Surprising as it may sound, firm estimates of groundwater potential of India are lacking at the moment. In 1960, Raghav Rao employed indirect methods of estimation, in which rainfall and soil percolation quality were used to assess groundwater recharge. He placed the groundwater potential of India at about 219 million acre-feet (26 mhm), with statewise breakup as follows on next page.

Ever since 1960, it is Raghav Rao's estimates which are being quoted – with some reservation – in many official and semi-official documents, e.g. the Report of the Second Irrigation Commission (1972), the Report of the Task Force on Groundwater Resources (Planning Commission, 1972), etc. However, many states - notably, Haryana, Punjab and UP – have expressed serious reservations about these estimates. They believe that direct methods of

	Million acre-feet
Andhra Pradesh	17.2
Assam region	16.7
Bihar	21.9
Delhi	0.3
Gujarat	10.2
Haryana	3.5
Himachal Pradesh	0.9
Jammu and Kashmir	4.0
Madhya Pradesh	26.7
Kerala	5.4
Tamil Nadu	11.5
Maharashtra	12.6
Karnataka	10.0
Punjab	6.9
Orissa	16.0
Rajasthan	3.4
Uttar Pradesh	35.5
West Bengal	16.1
All-India	218.8

groundwater estimation would yield higher estimates of groundwater potential. In fact, the Groundwater Investigation Directorate of Uttar Pradesh, using the 'water balance' method, placed the utilisable groundwater potential of Uttar Pradesh at 42.6 million acre-feet, as compared to 35.5 million acre-feet estimated by Raghav Rao. Moreover, taking a dynamic view of groundwater potential, the Directorate raised this potential to 49 million acre-feet. In this dynamic assessment, it took account of additional canal irrigation from Gandak, Ramganga and Sarda Sahayak projects which are under construction. For, seepage from canal and canal-irrigated fields are, besides rainfall, an important source of groundwater recharge. As Uttar Pradesh harnesses more and more of its surface water resources – estimated at 180 million acre-feet – its utilisable groundwater potential should also rise further. Whether it would be as high as 140 million acre-feet, which the Up Irrigation Department hopes to extract annually in order to support a crop intensity of 240 per cent in the entire state, is hard to say without more detailed information. Anyway, the pertinent point is that groundwater potential is not a static concept. Consequently, estimates of groundwater potential are liable to upward revision as more and more irrigated agriculture comes into vogue.

There are, indeed, very many interesting questions relating to groundwater utilisation and its economics. What is the crop pattern of areas irrigated by groundwater sources, and how does it compare with the crop pattern in canal-irrigated areas, or dry areas ? What are the capital and operating costs of various groundwater structures – such as open wells, dug-cum-bore wells, and tubewells; and how does irrigation cost of a structure vary with the source of energy used ? What are the internal scale economies in tubewell irrigation and how would cost of tubewell irrigation compare with effective resource cost of canal irrigation ? What are the external diseconomies of tubewell irrigation ? How much of the utilisable groundwater potential has been tapped so far and what are the factors which have promoted or inhibited the exploitation of groundwater resources in various regions of India ? In particular, what has been the contribution of rural electrification, land consolidation, and financial institutions in the rapid expansion of modern groundwater irrigation like tubewells and wells fitted with power pumps ?

Literature is gradually accumulating as scholars and investigators explore the above questions. A time is bound to come, soon, when as much knowledge about groundwater irrigation would be available as is currently available about canal irrigation. At the moment, it is difficult to present a worthwhile summary view of answers about the above questions.

3

Groundwater Estimates of India – A Note of Caution *

Over the years, I have written in this journal many times about developments in groundwater irrigation in India. Presently, my concern is with the problematics of our irrigation potential from our groundwater reserves, a topic that I have last discussed in considerable detail more than a decade ago (Dhawan 1977). The reason for 'revisiting' it now stems from the recent attempts by our groundwater agencies, notably the Central Groundwater Board, to what I fear is an overprojection of the irrigation potential of our groundwater resources. The Groundwater Board, which is the apex groundwater body in India, now claims that our groundwater resources can annually irrigate about 80 million ha[1] of crop area (not *land area*). I am somewhat intrigued by the fact that this number of 80 million ha is the same which B B Vohra, a popular protagonist of groundwater irrigation cause, had begun mentioning a year or so ago. He had, however, arrived at this number via a novel route. He assumed (by misreading my research work) that land productivity under groundwater-irrigated lands was twice that under surface water-irrigated lands. Therefore, the reported official 'ultimate' irrigation potential from groundwater resources of 40 million ha was multiplied by him by a factor of two. This hazardous 'production' approach for sizing up resources of groundwater totally ignores the linkage between groundwater and surface water; for, a good deal of groundwater is of canal/

**Economic and Political Weekly* (Forthcoming)

1. This areal number (abbreviated to 'mha') needs to be distinguished from a similar number expressed in volumetric terms, namely, million hectare-metres (often abbreviated to 'mham' or MHM in technical writings on water). Not infrequently, media writings do not distinguish between the two units of measurement, one being of area and the other being of volume.

tank origin. In Punjab, for example, half the groundwater raised by tubewells is the seeped-in water from the extensive canal irrigation works established at huge cost.

In what follows I shall state the reasons which sufficiently indicate the need for critical examination into the basis of this radical upward revision in the estimated groundwater potential.

In my earlier paper (Dhawan 1977) I had tried to underscore three things : (1) lack of firmness in the then available assessment of India's groundwater resources, (2) the imperative of distinguishing between stock and flow dimensions of groundwater resource, and (3) the likely enlargement or 'growth' of groundwater resource with the creation of more and more surface water-based irrigation works, especially canals. Therefore, given the tentative nature of the earlier estimates and the substantial development of surface irrigation works, one can neither reject the need for a revision in the groundwater estimates, nor be surprised by the upward direction of such a revision. But the magnitude of this revision by a factor of two – for some states by a factor of even three and more – certainly calls for a close look at the methodology under-lying the assessment. One cannot also overlook the possible inbuilt upward motivational biases in the new estimates, two of which might at once be suspected here. One stems from considerations of availing institutional credit for establishment of tubewells, dugwells, pumpsets, etc., while the other is linked with a sense of neglect by groundwater bodies and their personnel at the top in the matter of redressal of their needs of public resources/promotional avenues vis-a-vis canal-related bodies and their top personnel. The credit angle applies to state governments, always keen to keep the channels of bank finance open. Since the banks – notably the National Bank for Agricultural and Rural Development (NABARD) which is the apex body for disbursement of farm loans – shut off credit for new groundwater schemes in a block if 85 per cent of that block's groundwater potential is already tapped, state administrations as well as politicians are susceptible to inflating the estimated groundwater potential on the one hand and understating the existing level of groundwater utilisation (i.e. draft). Even Government of India may not be above such a predisposition to the extent that it gets external finance from World Bank for groundwater development : this aid helps in augmenting foreign squarely fixed on non-construction of drainage system within canal commands.[9] Thus the problem of waterlogging is not an inevitable consequence of dam building. It could be effectively checked with adequate provisions for drainage, whether through horizontal drains or through vertical drainage via wells and tubewells. In arid regions, the problem of soil salinity is best contained through the use of drip and sprinkler modes[10] of watering crops, because the disposal of drained saline water is an exceedingly tough problem in such areas.[11]

THE SEISMIC ANGLE

Another area of concern relates to damage from earthquake. There is an apprehension that the huge mass of water behind a dam may induce seismic disturbances in the neighbourhood area. Increased seismic activity in the Koyna region has been ascribed by some persons to the Koyna dam. This is an issue which can be best resolved by geologists and seismologists. What is of greater concern is when a dam is proposed on a dam-site situated in an already well-marked seismic belt of the earth, e.g., the proposed Tehri dam in the Garhwal Himalayas. In such cases, a major breach in a dam following an earthquake of severe intensity could be an instrument of utter disaster instead of economic development. Here, too, one has no choice but to go by the judgment of civil engineers as to how much earthquake-proofing can be achieved for dams built in seismic zones. If there is no unanimity among them on the new technology of rock-filled dams, scaling down the dam height would go a long way in lessening the anxiety about damage from dam burst due to an earthquake.

---

9. One of the main reasons why the drainage component has been missing from our major canal networks is the paucity of investment funds. After all, the cost of drainage network, both in capital and maintenance terms, is no different from that of the canal network itself. It is mainly because of this paucity of funds that even the full complement of the canal system is not completed on many irrigation works even after several years of their being commissioned.
10. Drippers and sprinklers are quite costly to install and operate, not to mention that these cannot be used for all crops. For example, drippers are economical for horticultural farming, while sprinklers cannot be used for irrigating paddy which absorbs almost half the irrigation water in India.
11. Soil scientists, while quoting favourable benefit-cost ratios for reclaiming salinised fields through their research based techniques, never reckon with the problems arising from draining saline waters into rivers that are already being used for irrigation and drinking water purposes.

result that the extent of upward revision in 'ultimate' irrigation potential from groundwater varies from state to state. However, there is a singular case of the CMP-ruled West Bengal whose revised ultimate irrigation potential is shown to be 25 per cent less than the earlier estimate of 2.5 million ha. This astounding downward reassessment for a state said to be floating on groundwater signals the need for a fuller probe into the procedures adopted for revising the groundwater estimates.

State-wise changes in groundwater irrigation potential are listed in Table 3.1, in which I have attempted to categorise the states by alluvial and rocky geology.[3] However, I wish to caution the reader that the two separate reports from which this table is built are not readily available. An air of secrecy, if not confidentiality, has come to surround such reports eversince the pernicious practice of affixing label 'For Official Use' on their cover page has come into vogue. All kinds of evasions are practised when you seek copies of these. Only in February 1989 I stumbled upon a copy of the report of the Minor Irrigation Group for the Seventh Plan prepared five years earlier when a colleague of mine, an erstwhile member of the Planning Commission, while sorting out his papers, passed it on to me. I was surprised because I had been told by the Irrigation Advisor to Planning Commission in 1988 (after I had become a member of the Steering Committee on Irrigation for the Eighth Plan) that only a summary of recommendations was received from 7th Plan Group on Minor Irrigation. Furthermore, despite being a member of the 8th Plan Group on Minor Irrigation, I have not been given the interim or final copy of this Group's Report till today (15th January 1990). Only a month ago I received a copy of this report that was routinely distributed to members of another high level Committee on instructions from its Chairman.

It will be noticed from Table 3.1 that the upward revision in the estimated groundwater potential is steeper in the case of groundwater-short states of the hard rock region (that is south of the Vindhyas) than in the Indo-Gangetic plains north of the Vindhyas. More specifically, the revised potential for the hard rock states is now over three times the pre-revised level.

3. The original data sources simply list groundwater potential by state and by Union Territory.

It was in November 1982 that the Government of India constituted the Groundwater Estimation Committee, which submitted its report in March 1984 (GEC, 1984). Anyone questioning the revised estimates of groundwater potential-which indicate a tremendous scope for establishing new wells in states said to be experiencing groundwater depletion - gets the stock reply, in seminars or elsewhere, that these are worked out according to the new methodology enunciated in the GEC report. If you asked for a copy of this report, you did not get one. For nearly two years I was repeatedly promised a copy by the erstwhile chairman of the Central Groundwater Board whose predecessor, Dr. B.D. Pathak, was also the chairman of the GEC.

The promises were orally made whenever we bumped into each other in seminars/workshops/symposia/conventions. The copy I finally got was, thanks to my becoming a member of a committee set up nearly three months ago under the chairmanship of Dr. M.S. Swaminathan to review the working of the Groundwater Board. Even prior to becoming a member of this Committee, I had expressed my misgivings about the revised estimates in a paper read at a workshop on groundwater at Coimbatore, held under the auspices of the Water Technology Centre, TANU. But I can be more definitive now in examining the re-estimational methodology.

Admittedly, I am not well qualified to appraise the methodology recommended by the GEC. Yet, it is instructive to peruse its report. One is immediately struck by the credit angle, namely, NABARD tapping the World Bank finance for groundwater development in India. In point of fact, the first reassessment of our groundwater potential was initiated by ARDC (now called NABARD) in November 1977 when it constituted Groundwater Overexploitation Committee. It was also chaired by the then chairman of the Groundwater Board.

Earlier, the norm for rainfall infiltration (upto the groundwater table) for hard rock areas was upto 7.5 per cent of rainfall of the area in pre-revised guidelines of 1972 issued by the then Central Ministry of Agriculture to all state governments and financial institutions. The Groundwater Over-exploitation Committee liberally lifted this norm to the range of 10-15 per cent without making any distinction in the nature of the rocks. The corresponding norm for sandy (alluvial) areas was placed at 20 to 25 per cent (15 to 20 per cent with large clayey content). The GEC recommended the following norms for various types of rocky areas :

I.	*Granitic Terrain* :	
	(a) Weathered and fractured	10 to 15 %
	(b) Unweathered	5 to 10 %
II.	*Basaltic Terrain* :	
	(a) Vesicular and jointed	10 to 15 %
	(b) Weathered	4 to 10 %
III.	Phyllites, limestones, sand stones, quartzites, shales etc.	3 to 10 %

With the establishment of state-wise groundwater investigation organisations since 1971, data on water table fluctuations began to be compiled, thereby permitting recourse to water level fluctuation instead of rainfall approach to groundwater recharge. In this direct method rise in water table in sample observation wells in an area during the main monsoon period is the basis for estimating volume of groundwater recharge. Multiplying this vertical rise by the area of a tract, say, a block, one obtains the estimated volume of soil/earth in whose pores and empty spaces (e.g. fractures in rocks) is stored the groundwater recharge from rainfall, canal seepage, etc. Multiplying this volume by "specific yield" factor (a sort of measure for the relative volume of water to the total volume of wet soil/earth), one ends up with the final measure for the groundwater recharge. The Groundwater Overexploitation Committee had earlier recommended the following values of "specific yield"[4] :

sandy alluvial areas	12 to 18 %
silty alluvial areas	6 to 12 %
granites	3 to 4 %
basalt	2 to 3 %

4. These comparative figures of "specific yield" broadly indicate the bountifulness of groundwater occurrence (per unit area of land) in alluvial vis-a-vis hard rock areas of the country.

The GEC recommended that the estimate of "specific yield" be obtained from detailed work done by different agencies through "pumping" tests. As a guide, however, it suggested the following set of values for this parameter :

	For rocky tracts	*specific yield (%)*
1.	granites	2 to 4
2.	basalts	1 to 3
3.	laterite	2 to 4
4.	weathered phyllites, shales schist and associated rocks	1 to 3
5.	sandstone	1 to 8
6.	limestone	3
7.	highly Karatified limestone	7
	For alluvial tracts	
8.	sandy alluvial area	12 to 18
9.	valley fills	10 to 14
10.	silty/clayey area	5 to 12

The new estimates of groundwater potential,[5] prepared as per the norms of the Groundwater Over-exploitation Committee of 1977, were brought by the Central Groundwater Board (and some states) to the notice of the 7th Plan Working Group on Minor Irrigation in 1984. But this Group did not accept these estimates and proceeded to make its 7th Plan recommendations on the basis of the old estimate of 40

5. The magnitude of these new estimates is not mentioned in the report (GOI, 1984). It is intriguing that nowhere (neither in the reports of the Plan Groups on Minor Irrigation nor in the original reports of the Central Groundwater Board and the state groundwater investigation organisations) one is given any idea as to the magnitude of revision being sought in the earlier estimates for each state. A discerning -- and determined -- reader would have to do so on his own by collating information from diverse sources. As a result of the new norms of the Groundwater Over-exploitation Committee, the volumetric estimate of gross groundwater recharge went up from 35 to 42 mham, and on to 45 mham in the wake of the GEC report (*Source* : Personal Communication from CGWB).

million ha. The Group also recorded in para 3.03 of its report (GOI, 1984) the following noteworthy observations :

"These (new estimates) were discussed at length and the Working Group concluded that before the revised figures are adopted, detailed criteria and calculations forming the basis of the revision would have to be presented and *scrutinised by a team of experts* on the subject. The Working Group recommends constitution of two separate groups of experts for surface water and groundwater to examine the revised figures in detail" (emphasis added by the author).

It was in May 1985 that instructions were sent by the Centre to the states to constitute expert groups to prepare revised estimates as per the methodology contained in the report of the Groundwater Estimation Committee of 1984. It was on the basis of these state-wise reports of expert groups that the Groundwater Board once again pressed the 8th Plan Working Group on Minor Irrigation to adopt the higher estimates of ultimate groundwater potential. While adopting the new estimates, the 8th Plan Working Group recorded also some reservations (enumerated in para 5.4 of its report).[6] In view of the reservations, the report makes the following commendation to the Ministry of Water Resources :

"Earlier Groundwater Estimate Committee was constituted in 1983-84. The Ministry may constitute another Committee this year so that its report is available by end of VIIth Plan period. The Committee may be headed by an expert having knowledge both in surface and groundwater hydrology". (p. 17).

It is disturbing that no scrutinisation of calculations of the revised estimates, as was pleaded by the 7th Plan Group on Minor Irrigation in 1984, has been carried out so far. Though the report of the 8th Plan Group on Minor Irrigation does not explicitly underline the need for such a scrutiny, yet the imperative of such a scrutiny cannot be brushed aside. It is most dissatisfying to be told that the new estimates have been endorsed by a meeting of the state secretaries

6. What transpired in the meetings of this Group, is not known to me, as I was not present in the later meetings of this Group, not having received any intimations, written or telephonic, about the missed meetings, nor even receiving their minutes afterwards.

of irrigation departments in June 1989. After all, even the representative of NABARD on this 8th Plan Group, a groundwater expert of considerable repute and research experience in groundwater balance studies, was far from reassured by the explanations of the CGWB on the revised estimates. He submitted his misgivings in writing and these are reproduced in the report (GOI, 1989). They underscore the need for scrutiny of the revised estimates of ultimate groundwater potential. For the benefit of concerned readers, I am reproducing, in an appendix to this paper, the entire letter of the NABARD representative on the new estimates.

It would seem that the 8th Plan Group on Minor Irrigation, unlike the 7th Plan Group, was under some compulsion to adopt the revised estimates of groundwater potential. One conceivable compulsion could be the impossibility of playing the "numbers game". By the end of the 7th Plan, the created irrigation potential from groundwater sources was feared to touch a level of around 35 million ha, leaving a balance of only 5 million ha to be developed in the subsequent 8th Plan period. As the target for additional groundwater potential for the 7th Plan period was 7.1 million ha and the achievement too was anticipated to be very near the target (near about 6.98 million ha), it would have been a climb down for the "planners" to target for groundwater development on a lower scale than before – more so when the then Prime Minister was announcing (on his own) ambitious programmes of establishing new tubewells and dugwells.[7] While in the case of Uttar Pradesh and Gujarat the likely achievements by the end of 7th Plan showed up as cases of 'piercing the ceiling', the situation was one of imminent 'bumping against the ceiling' in respect of states like Tamil Nadu, Bihar and Rajasthan if groundwater development programmes for these states were to be planned even at normal pace. Having lifted the ceiling, the Group recommended a bold target of additional 10 million ha of groundwater-based irrigation capacity for the 8th Plan period.

7. At one time the PM announced the construction of 5 million new dugwells and an equal number of new shallow (i.e. private) tubewells. These numbers could not be "accommodated", because (1) 9 out of 12 million number of ultimate feasible dugwells had already come to the established as per the progress reports submitted to the Planning Commission, and (2) we were expected to cross the 4 million mark of ultimate feasible shallow tubewells by a margin of 0.7 million by the end of the 7th Plan.

I should reiterate here that there is reason to expect upward change in the ultimate groundwater irrigation potential. The added recharge to groundwater resource from the ongoing development of canal irrigation is bound to enhance the volume of groundwater potential, more so when our canals are known to be much more leaky than visualised by their designers/builders (Dhawan 1989). Further, the previous estimate of 40 million ha was indeed based on a uniform irrigation application of 0.66 metres of groundwater all over the country (i.e., D = 0.66 m or 66 cm). For this country of continental diversity in soil-climatic conditions, such a uniform parameter is totally unwarranted. But the revision should not become an occasion for manipulating an over-projection of the groundwater irrigation potential. I am listing in Table 3.2 the new values of irrigation application, alongwith their range,[8] which are now being adopted for the revised estimates. Do these appear realistic ? How does one arrive at the magnitude of this crucial parameter ? What assumptions have been made about the crop pattern on groundwater-irrigated lands ? Is it realistic to assume irrigation requirements for a crop as per the recommendations arising out of researches done under farm research station conditions of tiny plots and measured doses of water ?

Finally, I would like to caution against overblowing our achievements in creating additional irrigation capacity. In this regard the method of arriving at estimated new irrigation capacity needs to be corrected. The Central Water Commission has already made a modest beginning in this regard in respect of created irrigation capacity from major and medium irrigation works based on surface water.[9] By removing elements of over-reporting of new irrigation potential created through major and medium schemes, the CWC had derated (downgraded) the achievements by about 2.4 million ha. The simplistic approach of multiplying the estimated number of wells established with the area norm per well calls for a check with ground

8. A word of explanation about the range values. Under the revised approach, estimates are first built for each block/watershed/taluka/district of a state. The range value is based on these 'micro' level data.

9. I am writing this on the basis of the Report of the Working Group on Major and Medium Irrigation Programme for the Eighth Plan (1990-95). See its page II-4 and Annexure II(1) for statewise derating of irrigation potential from major and medium irrigation works.

reality, notwithstanding the fact that the area norms are continually revised downwards from Plan to Plan (the area norms recommended by the Groundwater Estimation Committee are shown in Table 3.3). What is pernicious in this approach is that the resultant figure of additional irrigation potential in a Plan period is cumulated over time. That some of the new capacity is not additive with the earlier existing capacity is seldom rackoned with. Lakhs of dugwells have gone out of use after their owners installed tubewells in the Indo-Gangetic plains.

TABLE - 3.1

Pre-revised and Revised Estimates of Ultimate Irrigation Potential from Groundwater Resources.

S.No.	*State/Region*	*Pre-revised figure (000 ha)*	*Revised figure (000 ha)*	*Col.4 as a multiple of col.3*
1	*2*	*3*	*4*	*5*
I.	*States with rocky underneath*			
1.	Andhra Pradesh	2,200	5,190	2.4
2.	Gujarat	1,500	4,810	3.2
3.	Karnataka	1,200	3,120	2.6
4.	Madhya Pradesh	3,000	12,700	4.2
5.	Maharashtra	2,000	5,840	2.9
6.	Tamil Nadu	1,500	3,350	2.2
	Total 1 to 6	11,400	35,010	3.1
II.	*States with alluvial underneath*			
7.	Bihar	4,000	7,180	1.8
8.	Haryana	1,500	1,880	1.25
9.	Punjab	3,500	3,820	1.1
10.	Uttar Pradesh	12,000	18,000	1.5
11.	West Bengal	2,500	1,880	0.75
	Total 7 to 11	23,500	32,760	1.4

cont.

III.	*Other States*			
12.	Assam	700	1,560	2.2
13.	Himachal Pradesh	50	74	1.5
14.	Jammu & Kashmir	150	783	5.2
15.	Kerala	300	990	3.3
16.	Orissa	1,500	5,400	3.6
17.	Rajasthan	2,000	3,440	1.7
	All India	40,000	80,380	2.0

Source : Col. 3 is from Annex III of Government of India (1984), and col. 4 is from Annex 39 of Government of India (1989).

TABLE - 3.2

Irrigation Application per Crop Hectare Area Assumed in the Revised Ultimate Irrigation Potential from Groundwater.

S.No.	*State*	*Overall value of 'D' (cm)*		*Range of 'D' (if any) (cm)*
1.	Assam	128		-
2.	West Bengal	94		60-167
3.	Tamil Nadu	76		36-94
4.	Andhra Pradesh	71		56-91
5.	Kerala	69		-
6.	Maharashtra	55		40-75
7.	Karnataka	44		35-63(?)
8.	Madhya Pradesh	40		
9.	Orissa	37		34-44
10.	Punjab	40		
11.	Rajasthan	39		39-42
12.	Uttar Pradesh	36		
13.	Bihar	40		
14.	Gujarat	40	(unconfined aquifer)	32-50
		43	(confined aquifer)	36-50
15.	Haryana	39		
16.	Himachal Pradesh	39		
17	Jammu & Kashmir	47		39-60
18.	Delhi (UT)	39		
	All-India	47.6		

Source : Government of India (1989).

TABLE - 3.3

Area Norms for Different Types of Wells in Different States

Sl.No.	*Type of Minor Irrigation Work*	*Area Irrigated (ha.)*
1	*2*	*3*
1.	ANDHRA PRADESH	
	Dugwell with mhot	0.5
	Dugwell with Pump set	2.0
	Private Tubewell	4.0
2.	BIHAR	
	Dugwell without Pump	
	(i) Upto 3m dia	0.6
	(ii) From 3 to 6 m dia	1.0
	Dug cum borewell Tubewell	2.0
	(i) 10 cm dia	4.0
	(ii) 5 cm dia	2.0
	Diesel Pump set on Dugwell/ Surface water sources	
	(i) 5 HP Pump set	2.0
	(ii) Pump above 5 HP	4.0
3.	HARYANA	
	Dugwell	1.2
	Shallow Tubewell	4.3
4.	PUNJAB	
	Dug well	1.0
	Shallow Tubewell	5.0
5.	MADHYA PRADESH	
	Dug well	1.0
	Shallow Tubewell	6.8
6.	MAHARASHTRA	
	Dug well with Pump set	2.0
	Dug well with mhot	0.5
7.	TAMIL NADU	
	Private Tubewell	8.0

cont.

Table 3.3 cont....

1	*2*	*3*
	Filter Point	4.0
	Boring in well	0.8
	Deepening of well	0.8
8.	UTTAR PRADESH	
	Masonry well	1.0
	Persian wheel (addl.)	0.5
	Boring (small/marginal farmers) addl.	0.5
	Pump set on boring	5.0
	Tubewell	5.0
9.	TRIPURA	
	Shallow tubewell	4.0
	Artesian Well	0.5
10.	WEST BENGAL	
	Shallow Tubewell	3.0
	Dug well	0.4
11.	RAJASTHAN	
	Dugwell	2.0
	Low duty Tubewell	2.0
	Dug cum borewell	4.0

Source : GEC (1984).

APPENDIX

NATIONAL BANK FOR AGRICULTURE AND RURAL DEVELOPMENT

S.P. Sangal
Deputy General Manager
D.O. Tech No./1438/MI/88-89
February, 1989
4-3-1989

My dear Saksena

Kindly refer to the discussions held on 27th February, 1989 in connection with minor irrigation working group meeting for formulation of Eighth Five Year Plan. While discussing the strategy for groundwater development during the Eighth Plan period the conversion of the volumetric groundwater availability for irrigation to area basis was discussed. In this connection we had made certain observations which are again given below for your consideration.

(i) At present the net cropped area in the country is about 143 mha.
(ii) At present the Central Water Commission has estimated ultimate potential through Major, Medium and Minor irrigation works at 113 mha. Out of these 40 mha are through groundwater development, 15 mha through surface water minor irrigation works and 58 mha through major and medium surface irrigation projects.
(iii) The present estimates for ultimate minor irrigation potential through groundwater resources are 40 mha.

Keeping in view the above figures, if present estimate of 40 mha through groundwater development is increased to 80 mha as suggested in the agenda note, it will leave a balance of only 18mha as the ultimate feasible potential through major and medium irrigation projects unless the figure of 48 mha for these works is also suitably revised upwards.

We had therefore suggested that the figure of 80 mha through groundwater works as given in the agenda note may be examined by Group comprising of members from the Water Resources Division of CWC and the Ministry of Agriculture and others, as considered fit by the Chairman of the Working Group before this figure is finally accepted for incorporation in the working Group Report.

It was also suggested that the Ministry of Water Resources may bring out the complete report on the basis of which increase of ultimate minor irrigation groundwater potential has been suggested from the present 40 mha to 80 mha. If a copy of the same is available in the Ministry you may kindly send the same to the undersigned.

REFERENCES

Dhawan, B.D. (1977). India's Groundwater Resources, Review of Agriculture, *Economic and Political Weekly*, March issue.

Dhawan, B.D. (1989). Studies in Irrigation and Water Management. Commonwealth Publishers, New Delhi

Government of India (GOI, 1984). Report of the Working Group on Minor Irrigation for Formulation of the Seventh Plan Proposals for the Years 1985-90. Minor Irrigation Division, Ministry of Irrigation, New Delhi (mimeo).

Government of India (GOI, 1989). Report of the Working Group on Minor Irrigation for Formulation of the Eighth Plan (1990-95) proposals constituted by Planning Commission. Minor Irrigation Division, Ministry of Water Resources, New Delhi, July (printed but "for official use" only).

GEC (1984). Report of the Groundwater Estimation Committee : Groundwater Estimation Methodology. Ministry of Irrigation, Government of India, New Delhi (printed but unpriced).

4

Management of Groundwater Resource*

Direct versus Indirect Regulatory Mechanisms

Contrary to the general impression, the irrigation resources of India are far from bountiful. In point of fact, these are rather inadequate considering our resources of arable land, our large and steadily growing population, and the present state-of-the-arts of tapping water resources and field application of irrigation water. Even double cropping of 140 million ha of presently cultivated land with the help of irrigation appears to be a feasible proposition for only some regions like the East Gangetic plains (West Bengal, Bihar and East UP), Assam, Orissa, and the western coastal strip (from Surat downwards). Given the preference for heavily-irrigated crops like paddy and sugarcane, the irrigation resource would prove inadequate for double cropping in the rest of the country. For instance, in the vast Deccan plateau the available water resources would barely suffice for bringing 30 per cent of the crop area under irrigation–the percentage can possibly be doubled if sugarcane is not permitted under canal commands. Likewise, the rise of paddy cultivation in the wheat belt of north-western India (notably in Punjab and Haryana) is not conducive for double cropping of irrigated agriculture on a sustainable basis.[1]

In general, groundwater resource inherently tends to be much less plentiful than surface water.[2] This differential arises because a smaller fraction of rainfall tends to infiltrate down to the groundwater table than the fraction flowing into the rivers. To illustrate, in arid areas

*Reprinted from *Economic and Political Weekly* September 5-12, 1987

1. Though the water resources of Punjab are quite bountiful for its own requirements, these have to be shared with Haryana and Rajasthan under national imperatives.
2. Reference to groundwater resource here is in a flow, and not stock, context–stock of groundwater certainly exceeds surface water stock.

like western Rajasthan natural groundwater recharge amounts to about one per cent of the annual precipitation; in hard rock regions like the Deccan and the Bihar plateaus the percentage may be 5 to 9 of annual rainfall; and in flat, alluvial regions like the Gangetic and the Indus plains the percentage is about 20 to 22 of rainfall. On the other hand, surface run-off (that portion of rain water that fails to infiltrate into the ground) may range between 20 and 60 per cent of the precipitation depending upon the slope and the extent of vegetal cover of the land.[3] At the national level, the groundwater resource is reckoned at 26 million hectare-metres (one ha-metre = 10,000 cubic meteres = about 100 acre-inches) as against 66 million ha-metres of (utilisable) surface irrigation resource.[4] Given (a) this relative scarcity of groundwater vis-a-vis surface water under natural conditions, (b) the widely-shared belief that groundwater irrigation is superior to surface irrigation in the matter of raising land productivity,[5] and (c) groundwater can be conveniently tapped under individual enterprise, it is easy to visualise the emergence of pressure on groundwater resource, either in the shape of over-extraction or overcrowding of wells in a given aquifer (or both).

---

3. Besides groundwater recharge and surface run-off components of rainfall, a good deal of rainfall is absorbed by plants for sustaining dry farming, trees, grass lands, shrubs, etc.

4. The estimate of surface irrigation potential is likely to rise considerably in view of the current thinking to interlink some rivers.

5. This author's own empirical investigations revealed that in the late seventies each net hectare of irrigated area recorded the following levels of land productivity for different sources of irrigation :—

(tons)

	In Foodgrain Equivalent Units	
	AP and Tamil Nadu	*Punjab and Haryana*
Surface Irrigation		
(a) Tanks	2.0 - 2.3	NA
(b) Canals	2.6 - 3.4	2.4 - 3.2
Groundwater	5.7 - 6.5*	5.5 - 5.7**

* mainly dugwells.

** mainly tubewells.

PROBLEM

Overextraction in the specific context of groundwater resource means that withdrawals tend to exceed the annual replenishments via groundwater recharge, the excess being accounted for by a permanent reduction in the volume of groundwater stock underneath – it is in this sense that over-exploitation of groundwater is described as groundwater 'mining'.[6] Its unmistakable manifestation is the lowering of groundwater table. In the economics of groundwater irrigation, permanent recession in groundwater table raises both the capital and the operating costs of well irrigation. On the other hand, overcrowding gives rise to the problem of mutual interference among wells *when they are operated simultaneously*. In technical terms, the 'radii of influence' of wells overlap, that is, a well is located within the 'radius of influence' of another well. Due to mutual interference the operational cost of waterlifting rises : as the discharge of waterlifting mechanism diminishes, the duration for completing an irrigation operation rises.[7] Needless to add, the well owner with a smaller waterlifting capacity is disadvantaged more – in the extreme case he may not be able to lift any water till the well owner with the larger capacity stops the operation of his pump. In general, owners of traditional wells suffer when modern wells equipped with power-driven pumpsets come up in their vicinity.

MECHANISMS FOR GROUNDWATER REGULATION

Thus the task of groundwater regulation boils down to (1) controlling location of wells and (2) controlling groundwater draft by well owners so that their aggregate withdrawals in a year tend to equalise to a magnitude which ensures long-run stability of groundwater table. In short, groundwater regulation on the lines of industrial licensing is indicated whence some statutory control on location of a well and annual groundwater withdrawal by such a well

6. In arid regions, exploitation of groundwater is really like exploiting a mineral resource that gets exhausted in the end.

7. To illustrate with an example, if the rated discharge of a private tubewell falls from 10,000 to 7,500 gallons per hour due to mutual interference, time for applying irrigation to a depth of two inches to a one hectare field increases from about 11 to about 15 hours (about 34 per cent rise).

is exercised. In the Indian conditions, it is a stupendous task that is almost beyond the capacity of a state government. That may explain why Indian states have been lukewarm to the centre's model bill on groundwater regulation.

When landholdings are large (as in the south-western US), spacing between wells can be ensured by calling upon farmers to site their wells outside the radii of influence of their neighbours' wells. But when these are small and fragmented in space suitable distancing among wells cannot be ensured without denying permission to late-comers whose proposed wells may come within the radii of influence of the existing wells in an aquifer/watershed. This kind of groundwater regulation becomes questionable in the light of the fact that farmers with comparatively smaller landholdings are often the late-comers to an aquifer. In such a situation, mutual interference gets mitigated.

In contrast to the problem of mutual interference, the problem of overexploitation of groundwater resource is far more intractable--and with far more serious consequences in certain situations. If decline in water table in a tract continues unabated, the rising cost of groundwater irrigation may prove inadequate to arrest the process of decline in water table. In fact, even before the depth to water table reaches a level where marginal cost of exploitation equals the marginal benefit from a unit of water, groundwater resource may be damaged due to intrusion of saline/brackish waters from the neighbourhood. This is a real danger in shallow aquifers near the sea coast and in semi-arid regions underlain with brackish waters unfit for farming.[8] Since well owners seldom subject the pumped water to periodic check-up for salt content analysis, they come to learn of the intrusion of saline water into their aquifer rather late (usually when white crusts appear in the fields). In a way, rural electrification has enhanced pressure on groundwater resource. To begin with, electric pumpsets are best

8. In some semi-arid tracts one may encounter the phenomenon of layers of sweet water interspersed with layers of brackish water (e.g, in Mehsana district in Gujarat).

suited for continuous pumping of water from any depth.[9] Secondly, electric power for agricultural pumpsets is heavily under-priced both in comparison to its resource cost and the alternative sources of energy for lifting water. Thirdly, the linking of power tariff with horsepower of a motor instead of actual consumption of power in many states makes, well owners behave as *if the marginal cost of pumping water is zero*.[10]Thus the structure of power tariff, as also the pitch of electricity price, badly needs due modification. (Admittedly, need for rectifying these weaknesses in the electricity pricing have been earlier emphasised by other scholars).

Attempts to control both the density of wells and the well-spacings in a tract through indirect means have proved rather ineffectual. Linking of the grant of electricity connection for a pumpset with a prior certification that there is no other pumpset/well within a specified distance has been overcome by farmers procuring the needed certificate through bribing of officials – failing this they could opt for the next best solution,namely, the installation of a diesel engine. Likewise, stoppage of institutional finance when a tract has witnessed over-development of well irrigation may have slowed down further overexploitation of groundwater resource provided benefit-cost ratio based on funds drawn from non-institutional source is not found to be favourable by farmers in that tract. The trouble with this instrument, however, is two-fold: (1) coping with strong political pressures that prevent declaring a tract ineligible for fresh financing for well irrigation, and (2) its invocation rather late in the day by which time overexploitation has become a reality. Even when no new wells are coming into existence, the existing well owners tend to depress the water table as they enhance their withdrawals by gradually improving intensity of cropping, and altering crop pattern in favour of more resource-using crops which usually need more irrigation. It is only when some of them begin investing in water-economising devices (sprinklers, drippers, lining or piping of channels, etc.) that downward pressure on groundwater table may begin to ease.

9. A manually, operated waterlift becomes dysfunctional when depth to water table exceeds 15 feet. The same limit for an animal-operated, persian wheel is around 35 feet, and about 50 feet, and about 50 feet for 'mhot', a traditional animal-operated lift of southern, western and northern India. A diesel pumpset is only a shade less than an electric pumpset in coping with any depth to water table.

10. Tushaar Shah is an exponent of this view and has been pleading for reduction in electricity tariff in states like Gujarat so as to facilitate the emergence of markets for groundwater.

Short of total exploitation of groundwater resource under governmental aegis, one sees no ideal solution to the twin-problem of overexploitation and overcrowding in groundwater utilisation. In point of fact, some scholars (including this author) subscribe to the view that groundwater exploitation in the east Gangetic plains through public tubewells instead of individually-owned wells is a superior option given the preponderance of small and highly-fragmented holdings for which individually-owned wells are not a viable proposition. However, this is not a region for which one should apprehend the prospect of overexploitation of groundwater resource—even if the water table recedes, it is likely to stabilise, without any danger of intrusion of saline waters, at a new level not much below the river-bed level. Even for much of the western portion of the Indo-Gangetic alluvium, which has witnessed an explosive growth of individually-owned tubewells during the last two decades, one need not worry unduly about overexploitation of groundwater. Extensive development of surface irrigation works massively augments groundwater recharge in this region, not to mention that availability of surface irrigation *ceteris paribus* reduces need for groundwater irrigation. The existence of an extensive canal network and some perennially flowing rivers (fed by snow melt and water stored in the mountains of the Himalayan, Aravalli and Vindhyan ranges) would tend to prevent decline in water table below the river-bed level. Here the danger is in areas adjoining Rajasthan state (south-west UP plains, south-western Haryana, and southern Punjab) because of high occurrence of saline bodies of groundwater.

Outside the Indo-Gangetic plains, scope for direct exploitation of groundwater resource under state ownership is rather meagre. To begin with, groundwater formations here are mainly of the rocky type in which tubewell technology can play little role. Secondly, wherever tubewell technology is feasible (e.g., in the coastal alluvium), large size tubewells are not a feasible proposition at very many places. Thus it is not a worth-while proposition for the state to operate under its aegis myriad dugwells. Yet, it is here that groundwater resource is under great pressure. This pressure arises because groundwater endowment is poor. Though water in relation to land resource is comparatively more scarce here than in the Indo-Gangetic plains, surface water resources *vis-a-vis* groundwater resources are more abundantly available.

Wisdom lies in paying more attention to the development of surface irrigation works, so that a sizeable fraction of surface waters end up in groundwater table. But for such artificial recharge of groundwater table through extensive development of canal network in Punjab, Haryana and west UP intensive farming based on tubewell irrigation would have been decisively less than the high level attained by it now.

CONCLUDING REMARKS

The role of people's participation in water-shed management in areas of undulating topography and hill regions is quite well recognised, if not firmly established, in India. The problems of groundwater management are primarily those of plains regions where groundwater-based agriculture comes into vogue. It is not yet clear how people's participation can help in the resolution of the twin problem of overexploitation and over-crowding in groundwater irrigation in the vulnerable regions. In the matter of major irrigation works, the action belongs to the state. It is with regard to minor surface irrigation works that people's participation at the local level may prove helpful, thereby improving the fraction of rainfall infiltrating into the ground. In particular, community effort in properly maintaining irrigation tanks (desilting of tank bed; periodic strengthening of tank bund; proper maintenance of feeder channels, if any, etc.) is good for recharging groundwater as well. Likewise, percolation tanks can be taken up through community effort as well, provided good geological advice is forthcoming-otherwise these may become simply 'evaporation' tanks.

5

Economics of Tubewell Irrigation

with Special Reference to India*

Tubewell technology for tapping groundwater, though known for over a century, failed to diffuse notably in the Indian sub-continent till the advent of another technology, namely, the HYV seeds. These two technologies, acting concomitantly, have played a critical role in the agricultural transformation in India since the mid-sixties. In point of fact, spatial variations in the rates of farm growth in India are found by some scholars to be well correlated with the expansion in private tubewell irrigation. Outside the hard rock regions where tubewell mode of tapping groundwater for irrigation purpose is technically an infeasible proposition, tubewells have gradually displaced dugwells mode of irrigation has almost been given up by the farmers in the western Indo-Gangetic alluvium, notably in the two Punjabs of India and Pakistan which have prominently figured in the literature as note-worthy stories of green revolution of South Asia.

Because of the constraints of space and time, the author has per force to be selective in scope in his presentation here, as also skip over the details with regard to his methods that underlie the empirical results quoted here. However, an attempt has been made to list the references from which the interested scholars can obtain further details. The scope of exposition is restricted to the following aspects mainly :

---

* This paper was written at the invitation of the Afro-Asian Rural Reconstruction Organisation (AARRO), New Delhi for their International Workshop on Planning and Management of Minor Irrigation Schemes, held at Port Louis in Mauitius during 25-29 August 1986. It was briefly reported in the AARO Report of this workshop in its publication *Planning and Management of Minor Irrigation Schemes* (AARRO, New Delhi, July 1987, pp. 41-42). The full text appeared in *Journal of Soil and Water Conservation,* Vol. 32, Nos. 1 and 2, March and June 1988, as also in *Financing Agriculture,* Vol. 20 (1), January-March 1988.

1. Scale economies in tubewell irrigation;
2. External diseconomies of tubewell irrigation;
3. Role of the state;
4. Tubewell impact on the economy of the farmer.

SCALE ECONOMIES

Modern industry is characterized by substantial scale economies of production. However, this is not so for the activity of tubewell irrigation. While one certainly does not encounter a U-shaped cost curve over a considerable range of tubewell size(discharge rate varying from 0.014 to 0.168 m^3/second), scale economies tend to vanish with increase in tubewell size. In other words, a sort of L-shaped cost curve tends to emerge. Substantial scale economies are observed in only two components of capital cost of tubewell, namely, tubewell proper and the pumpset. But these tend to be offset on two counts. One, the capital outlay on water distribution system, which tends to rise relative to the outlay on groundwater extraction system as the discharge rate is raised, rises slightly more than the increase in tubewell size. Two, the pumping costs tend to rise faster than tubewell size as tubewell with large discharge rate entails *ceteris paribus* greater lift of water. Using a summary measure of scale economy, as reflected by the exponent 'b' in the cost function.

$$C = a \cdot X^b$$

where C is total cost;
X is tubewell size

the magnitude of 'b' is found to be around 0.80 over the tubewell size range 0.014 to 0.168m^3 /second (i.e.0.5 to 6.0 cusec capacity) for government-owned tubewells in Uttar Pradesh.[1] On the other hand, for individually-owned, shallow tubewells, ranging in discharge capacity from 0.3 to 2.0 cusec, the scale economies are indeed of somewhat greater magnitude (b = 0.56 to 0.71)[2].

As far as public tubewells are concerned, setting them up with capacity above 2.5 cusec may not be prudent on three counts. First,

1. For more details about this estimate, see Dhawan (1978)
2. Source : Dhawan (1982), Table 4.5, p. 80.

the proportion of transit losses of tubewell water rise with the increase in command area consequent upon increase in tubewell capacity. This means an increasing hiatus between the volume of groundwater lifted and the volume actually delivered to the crops. Second, the management task of attaining better timeliness in irrigation for beneficiary farmers becomes more onerous as the number of beneficiaries rise *pari passu* with the size of a tubewell. Third, the danger of external diseconomies of tubewell technology is somewhat linked with the discharge rate of a tubewell, as we shall show later in the paper. Of these three constraining considerations, the first and the second one are indeed not germane to the case of augmentation tubewells that are established to directly augment canal/tank water supplies to the farmers.

As far the optimum size of a tubewell for an individual farmers, the crucial determinant is his farm size. Tubewell being an indivisible investment, it proves uneconomic for farm holdings below a given acreage.[3] The indivisibility is accentuated if the land holding is fragmented in space, not an uncommon feature in the Asian agriculture. If a farmer can overcome his disability of small farm size, or fragmentaction of holding, by selling surplus tubewell water to his neighbour farmers, the viability of investment in a tubewell is certainly enhanced. This very scope for selling water to others can be an incentive for a large farmer to invest in a bigger size tubewell, and thus exploit the handsome scale economies that exist at the lower spectrum of tubewell size range.

EXTERNAL DISECONOMY

No blessing is an unmixed one. Whereas introduction of canal irrigation tends to give rise to the twin problem of waterloggingand soil salinization, development of groundwater irrigation with the help of power pumpsets tends to deplete – or even damage – groundwater resource. Since the goundwater extraction capacity of a tubewell is

3. The author has assessed minimum size of farm holding in the various regions of the Indo-Gangetic plains for which investment in an individually-owned tubewell was uneconomic even under very efficient conditions of farming. See Dhawan (1982), pp. 52-59.

several times that of a traditional waterlift[4] mounted on a dugwell, development of tubewell-irrigated agriculture tends to give rise to excessive exploitation of groundwater resource, with the result that the groundwater table may be lowered permanently. Of course, the extent of recession in water table is governed by numerous factors such as density of tubewells; their discharge rates and annual running hours; rainfall and its infiltration rate into the soil; extent of artificial recharge of groundwater through canal/tank seepage; availability of surface irrigation for meeting part of the irrigators' demand for water etc.

As depth to water table plays a vital role in the economics of groundwater irrigation, a decline in water table has serious implications. To begin with, a permanent lowering of water table by tubewell owners adds to the cost of tapping groundwater for all farmers who are late comers to the aquifer. The depth-elasticity, which is a measure of increase in cost as depth to water table increases, is of the order of 0.29 for a tubewell but is almost twice as much in the case of traditional wells equipped with animal-drawn waterlifts.[5] Leaving aside capital costs, tubewell installation, if accompanied by a permanent lowering of water Table implies substantial addition to operational costs of existing dugwell - owners the estimated depth-elasticity in respect of operational costs varies from 0.57 in the case of persian wheels and 1.09 in the case of 'charsa' (rope-and-bucket lift, also known as 'mhot' in peninsular India). More often, a decline in water table leads to reduced output of groundwater from open wells, thereby adversely impinging on the quantum of crop output of their owners. Unless the recession in water table is too deep to be coped with by deepening of the open wells, reduction in crop output may be avoided by extra expenditure on well deepening.

---

4. A comparative idea about the water extracting capacities of animal-operated and pumpset-operated wells can be obtained from the following mean values of discharge rates for a sample of private wells in Punjab:

   Animal-drawn :

i. Persian wheel	0.125 cusec
ii. 'charsa'	0.138 cusec

   Tubewell :

i. electric	0.552 cusec
ii. diesel	0.749 cusec

   Source : Dhawan (1982), Chapter 4.

5. Source : Dhawan (1982)

Even if there is no permanent decline in water table in the wake of tubewell development, yet some untoward consequences may arise for traditional well owners. The source of the trouble is the large but temporary 'drawndown' (i.e. recession) in water table when a large capacity tubewell is operating. As a result of this, functioning of other wells (including low capacity tubewells) in the vicinity can be adversely affected. These short-run external diseconomies, as distinguished from long-run external diseconomies arising from permanent recession of water table, can assume serious proportions when several large capacity tubewells are established and operated in unison. This is what happened when Haryana installed a large number of big tubewells in Karnal district to augment canal supplies going to its distant Hissar district.[6]

GOVERNMENT CONTROL

How do we avert the external diseconomies of tubewell irrigation? An obvious solution to the short-run externalities is proper spacing among wells so as to avoid mutual interference in tapping groundwater from a common aquifer. In other words, wells should be spatially so sited that each is outside the 'radius of influence' of the other. This requires some sort of governmental regulation on private decisions to establish irrigation wells. As regards the problem of long-run externality, the need for keeping groundwater withdrawals and groundwater recharge in long-run balance is all the more imperative. To put it crudely, we need a sort of licensing system, whereby each owner of a groundwater structure is required not to exceed his groundwater use beyond a specified limit. It is a tremendous challenge in administration, more so for nations with multitudes of farmers owning small parcels of land. No wonder, no noteworthy progress has been made in this direction in India, despite reports of serious depletion in water table in the wake of explosive growth of tubewells in certain tracts of the country. In the meantime, prudence demands that exclusive reliance on tubewell as a means of irrigation is avoided. Efforts should be made to introduce surface irrigation in tubewell tracts. The conjunctive use of surface and ground waters would be doubly helpful in stabilising water table. On the one hand,

6. For further details, see Dhawan (1975, 1982).

as part of the demand for irrigation is met from canal/tank water there would be lesser pressure on groundwater resources. On the other hand, seepage of canal/tank water would enhance groundwater availability.

Besides regulatory role, the state is also in a position to establish tubewells under its own aegis. In other words, it can earmark the exploitation of groundwater within public domain, whereby public tubewells, like public canals, meet farmers' irrigation needs. This is an attractive option for groundwater exploitation from many angles. Firstly, some of the scale economies in tubewell irrigation can be realised by setting up fairly large-sized tubewells. Secondly, irrigation needs of marginal and small farmers, whose land holdings are too small to permit the installation of individually-owned wells, can be easily catered to through public tubewells. They can thus contribute to the developmental process by switching over from rain-fed to irrigated farming. Thirdly, wherever the task of land consolidation proves intractable, the matter of establishing--and fully utilizing the rated capacity--their own tubewells is substantially eliminated in a public tubewell irrigation system. Fourthly, the problem of groundwater management (distance between wells, groundwater withdrawal per tubewell, etc.) becomes quite simple and manageable.

But these advantages of groundwater exploitation through government-owned tubewells have to be duly weighed against some disadvantages. Firstly, with reduction in farmer control over the source of irrigation, the productivity of groundwater is bound to be less on lands receiving irrigation from public tubewells than from their own tubewells. Secondly, in the absence of efficient management of public tubewells, farmers' income and production gains from irrigation can become quite low and uncertain in comparison to private tubewell irrigation situation. Therefore, governments which are not confident of managing public tubewell systems efficiently should better keep out of this option. Instead, they should encourage(a) sale of surplus tubewell water by farmers to their neighbours and (b) development of cooperatively owned tubewells that may be initially fostered by voluntary agencies. [7]

7. For some glimpses of success in this area, see Pant and Rai (1985).

IMPACT ON FARMER ECONOMY

It is difficult to offer a definitive account of the impact of tubewell on the farm economy. For one reason, there are so many myraid aspects of such an impact that one can barely enumerate these here in a section of a small paper. Another difficulty is one of isolating the contribution of tubewell investment *per se* from the impact of other factors and forces. For, the joint impact of tubewell and other factors/forces is the outcome of a much more interactive than a purely additive process. To the extent these other factors are structurally related to the very advent of tubewell irrigation (in the sense that tubewell is a pre-requisite to their use e.g. chemical fertilizers are not notably used by farmers without irrigation in low and medium rainfall areas in India), one may be justified in ascribing the entire impact to tubewell factor alone. More often than not, scholars take recourse to this stratagem when they view the entire differential in a farm parameter, say crop yield difference between canal-irrigated and rainfed lands, to be due to canal irrigation. Obviously, the assessment of the impact of a given type of irrigation can give rise to a wide spectrum of results over time and space depending upon the extent of conjunction occurring between irrigation and other factors. Since this conjunction weakens as one moves eastwards in the Indo-Gangetic plains, one finds that the impact of a private tubewell on a farmer's income/productivity of land diminishes in force as one moves from Punjab/Haryana states to Bihar and West Bengal states.

As between different sources of irrigation, an individually-owned tubewell has come to occupy a pre-eminent position. Being a perennial source of water, a tubewell permits crop activity throughout the year, more so in rainless and rain-deficient seasons of the year. This provides gainful on-farm employment both for farmer's family labour and landless, agricultural labourers throughout the year, thereby curbing both seasonal and irreversible out-migration to the metropolis in search of employment. Tubewell being both a perennial and dependable source of irrigation, it permits intensive use of land resource as the farmer is quite insulated against the production risk emanating from failure in rainfall. In point of fact, the intrinsic dependability of tubewell during a year of drought is so high that an owner can flexibly pump out extra amount of water to cope with enhanced water needs of the plants during drought conditions.[8]

As per the author‘s econometric research work based on the official data of the states of Punjab and Haryana, it is found that each net irrigated hectare brought under private tubewell irrigation in the later part of the seventies enhanced crop output to the tune of 5 tons of foodgrains which is more than twice the corresponding output impact of canal irrigation.[9] A similar differential between the two sources of irrigation emerges when the comparsion is conducted on the basis of gross irrigated hectare.

SUMMARY

Optimal size of a public tubewell is around 2.5 cusec capacity. Beyond this size, the scale economies are marginal, and not worth exploiting because of inter alia adverse external effects on other users of groundwater, a common property resource. As regards the optimum size for an individual farmer, he is constrained in taking advantage of the scale economies at the lower spectrum of tubewell size range by his farm size, its extent of fragmentation in space, and the scope for agri-business through sale of water to other farmers.

Groundwater extraction capacity of a tubewell being much more than that of a traditional well, tubewell irrigation generates a tendency of overdraft on groundwater aquifer, giving rise to a downward pressure on water table. Depth to water table being a major determinant in groundwater economics, a continual decline in water table has serious implications. Also, proximity in location proves detrimental in the short-run,especially for a well owner having a lower capacity waterlift. To cope with these external effects of tubewell irrigation, groundwater regulations have been suggested by hydrogeologists. To administer these regulations effectively is, however, a practical problem in countries with small land holdings. While this author is not very hopeful of the legislative efforts in groundwater regulation,

8. This intrinsic dependability in the event of a drought, arising as it does from the invulnerability of deeper aquifers of a tubewell to reduced infiltration due to drought, may be affected if the hydel source of electric supply for tubewells is adversely affected by drought.
9. Source : Dhawan (1986)

he would advocate that excessive reliance on tubewell irrigation be avoided. Instead, the reliance be placed on conjunctive exploitation of ground and surface waters. Exploitation of groundwater directly under state auspices is recommended if (a) the state is confident of managing the public tubewell system efficiently and (b) there is preponderance of small and fragmented holdings.

Being a perennial source of irrigation, tubewell encourages crop activity throughout the year. Its dependability during drought permits crop intensification of a high level as the farmer is insulated substantially against production risk. With the state providing assurance of price stability, owners of tubewells are well placed to a steadily rising standard of living.

REFERENCES

1. Dhawan, B.D. (1986). *Studies in Irrigation Impact : Issues of Productivity, Stability and Equity*, Institute of Economic Growth, Delhi (mineo; likely to be published as a book in 1987).
2. ____________ (1982). *Development of Tubewell Irrigation in India*, Agricole Publishing Academy, New Delhi.
3. ____________ (1978). "Scale Economies in Tubewell Irrigation", *Indian Journal of Agricultural Economics*, January-March, vol. 33, no. 1.
4. ____________ (1975). "Externalities of New Groundwater Technology on Small Farmers", *Indian Journal of Agricultural Economics,* July-September, vol. 30, no. 3.
5. Pant, Niranjan and Rai, R.P. (1985). *Community Tubewell and Agricultural Development,* Ashish Publishing House, New Delhi.

6

Scale Economies in Tubewell Irrigation*

This study is motivated by the practical question : What would be the optimum size of a tubewell if the vast groundwater resources of the Indo-Gangetic plains are to be exploited for public irrigation? Whether developing groundwater irrigation under public aegis is a superior option compared to the option of private ownership of means of groundwater irrigation is indeed a controversial issue. Though this issue is beyond the scope of this paper, two points in this connection may be noted.

Firstly, an understanding of the scale economies internal to a tubewell and its potential external diseconomies on others is of no small consequence for resolution of the controversial issue. That a tubewell can impose external diseconomies on other users of groundwater--and even on users of stream flows–is known.[1] Since the magnitude of external diseconomies of tubewell irrigation can vary positively with the size of a tubewell, one has to reckon with the trade-off between the (internal) scale economies and the external diseconomies of tubewell size can give rise to results that may be far from the optimum from the larger, social viewpoint.

Secondly, the case of public instead of private tubewells is much stronger for the east Gangetic plains of East Uttar Pradesh, Bihar and West Bengal than for the western plains of Indus and the Ganges in Punjab, Haryana and West Uttar Pradesh. For, the structure of land holdings and objective factors influencing investment in a tubewell in

---

* Reprinted from *Indian Journal of Agricultural Economics,* Vol. 33, No. 1, January-March 1978

1. For an exposition of external diseconomies of tubewell irrigation, see Dhawan (4) and (5). (Figures in parentheses denote references cited at the end of the paper).

these two regions are such that the bulk of the holdings in the eastern Gangetic plains, unlike in the western plains, cannot afford installation of private tubewells.[2] Thus, in the absence of public tubewells the vast groundwater resources of the eastern plains are likely to remain largely unutilized. In point of fact, Hanumantha Rao has forcefully pleaded that "public tubewells may continue to be the major source for the exploitation of groundwater potential in this region."[3]

Public tubewells can be deployed for irrigation in two alternative ways : (1) each tubewell with its own separate irrigation command area and (2) tubewells for augmenting water supplies in a government canal system. While public tubewells are predominantly used at present as 'command' tubewells, augmentation tubewells are being increasingly planned, both as a means for enlarging water supplies in a canal system and as an anti-waterlogging measure in canal commands with poorly drained land.[4]

SCOPE OF THE STUDY

The study is limited to assessing purely internal scale economies for tubewells with independent command areas, though some implications of the study for the development of augmentation tubewells are also indicated. The size range of tubewells studied is from 0.014 to 0.168 cubic-metre discharge per second, *i.e.,* 0.5 to 6.0 cubic-feet per second ('cusec'). While the least size considered here is about one-third the size of most public tubewells in the country (especially Uttar Pradesh), the largest size considered is probably nowhere in operation at the moment in the entire Indo-Gangetic plains. While an analysis of tubewells of size above 6 cusec will be meaningful, the non-availability of data prevents us from extending the exercise to such tubewells.

---

2. See Dhawan (6).
3. See Hanumantha Rao (9). A similar plea is made by this author for East Uttar Pradesh in (7).
4. This anti-waterlogging function of public tubewells has not received due attention in the controversy about public versus private tubewell. While all tubewells do play an important conjunctive role in optimal utilization of surface and groundwaters, it is the lack of prospect for intensive utilization of seeped water from cannal irrigation in poorly drained lands that demands the installation of tubewells on public account.

The outline of the paper is as follows : The availability of data, which has considerable bearing on the methods to be adopted in an analysis of scale economies, is taken up first. It is then followed by an exposition on methodology used in this paper. Next, the salient technical characteristics of the set of tubewells analysed are briefly spelt out, followed by a discussion about the cost structure of a tubewell. Tubewell being a capital-intensive activity, the scale economies in respect of capital cost are analysed in detail, followed by a brief analysis of the scale economies in operating costs. The results about the overall scale economies in tubewell irrigation are presented, followed by some policy implications. To the extent one can envisage that the underlying assumptions with regard to the geological characteristics of this study (e.g., the assumption that 50 per cent of the bore depth is water-bearing) are fulfilled in the Gangetic plains of Bihar and West Bengal, the results have relevance for the entire east Gangetic plains.

THE DATA

Assessment of scale economies is known to be beset with problems, of which the foremost is the data problem. To begin with, firms (here tubewells) of varying capacities may not exist. The common, or rather the standard size for public tubewells in Uttar Pradesh, has been till recently 1.5 cusec ever since the mid-thirties when public tubewell irrigation commenced in this pioneer State. Secondly, cost information about the few tubewells of capacity greater than 1.5 cusec is not readily available. Even if such data are gathered from the unpublished records of the State governments, these can pose serious problems of separating pure scale effect from other sizeable effects attributable to inter-tubewell variations in respect to technological, historical (age and prices) and aquifer characteristics.

Thus pure scale economies are better revealed by hypothetical cost estimates of tubewells of various sizes, provided these are underpinned with experience and expertise. Such hypothetical estimates are available in a technical memorandum of the U.P. Irrigation Research Institute (UPIRI).[5] Notwithstanding the fact that actual installation costs of irrigation works are notoriously much different from the

5. Chawla, et al. (2). But for this source, the present study would have been inconceivable.

engineer's hypothetical estimates, the credentials of the UPIRI are high in the irrigation world.[6]

However, a few modifications on the UPIRI data have been made in this paper. Firstly, while accepting the UPIRI estimates of capital cost of tubewells, cost on account of electricity transmission segment that is often borne by the State Electricity Board instead of the Tubewell Irrigation Department, has been duly taken into account. This macro view of capital cost in tubewell irrigation is needed in order to unearth true scale economies in respect of the scarce capital resource of the Indian economy. Secondly, annual operational cost on account of electricity consumption in the running of a tubewell has been assessed both on the basis of unavoidable power tariff now in force in Uttar Pradesh and the avoidable power tariff that is in direct proportion to annual electricity consumption. Thirdly, the UPIRI estimate of annual operational cost of a tubewell is firmed up by taking due account of other operational costs, e.g., salary of tubewell operator; expenditure on overhead tubewell organization; and expenditure on tubewell repair.

METHODOLOGY

Unit cost of tubewell irrigation is obtained by dividing annualized cost of a tubewell by its annual water output. While annualizing capital expenditure on a tubewell, one needs information about the useful life of a tubewell, which is placed at 20 years in the UPIRI report. However, it is important to note that different components of a public tubewell have different life spans. Whereas the pumpset and the accessories have a life of about ten years, the well assembly and the pump house may last for 20 years. Likewise, the life of the water distribution system is generally reckoned at 50 years.[8] A similar life can be assumed for electric transmission segment. In this paper, the following rates are used to annualize :

(1) electric transmission/water distribution system ... 2 per cent,

6. Of course, the engineers' expertise and experience are some among the numerous factors creating this disagreement. The other noteworthy factors are : deliberate under-statement of project costs so as to get go-head sanctions for a project; alterations in project design; price inflation; leakages in funds and materials; and increae in gestation period due to uncertainty about the availability of funds.
7. Actual replacement age for reconstructing an old tubewell is found to be about 18 years in Uttar Pradesh.
8. Source : Jain (8).

(2) tubewell assembly and pump house ... 5 per cent,
(3) pumping set ... 10 per cent.

As regards annual output of water per tubewell, it is determined primarily by the crucial variable annual running hours (H), which, when multiplied by the observed hourly water discharge of a tubewell, yields the needed information about water output per tubewell. A variety of factors, some on the demand side and some on the supply side, determine the value of 'H'.[9] It is assumed that the determinants of 'H' are neutral to the size of the tubewell. Thus, in the event of electricity rationing by the electricity authority the adverse effect on runnings hours of a tubewell is in no way related to its size. Likewise, the crop pattern in the tubewell command area is envisaged to remain unaffected by the size of the tubewell.

Evidently, if all the annual costs of a tubewell are unavoidable, the analysis of scale economies is unaffected by the value chosen for 'H'. A proximate case of this type arises when one part, flat electricity tariff (in direct proportion to horse power of the tubewell motor) is in operation, as is the situation today in Uttar Pradesh. If one or more costs are avoidable, the value of 'H' chosen does impinge on the scale analysis. In this situation, unit cost of tubewell water is sensitive to the 'H'. In order to see how far our results are sensitive to the choice of 'H', it is proposed to carry out the analysis for different values of of 'H'. There are 8,760 hours in a year, and the chosenvaluesof 'H' are : 2,000, 3,000, 4,000, and 5,000 hours.

An attempt has been made to summarise individual sources of scale economies (or diseconomies) by fitting the following regression equation to the cost data :

$$C = a.x^b$$

where C is the cost of a tubewell item;
x is a measure of the tubewell size.

If 'b' is one, constant returns to scale exist. If 'b' is less than one, scale economies are in operation over the range of tubewell sizes considered. Likewise, 'b' greater than one is indicative of scale diseconomies. In terms of scale analysis, (1-b) is a measure of scale economies/diseconomies : (1–b) > 0 is a measure of scale economy;

9. Demand factors influencing the running hours of public tubewells have been investigated in Dhawan (3).

(1-- b) = 0 is a situation of no scale economy; and (1– b) < 0 is a measure of scale diseconomy. However, this elegant approach to evaluating the nature and magnitude of scale economies is unsuitable in cases where scale economies exist upto a point, followed by either diseconomies or constancy of economies. Cubic functions, like $C = \alpha X + \beta X^2 + \gamma X^3$, would take care of a situation in which scale economies are followed by diseconomies, and their first derivative, when equated to zero, yields the point of inflexion where economies turn into diseconomies.[10] In case scale economies follow an L-shaped path--as has been found in some studies on industrial activities–, no elegant function for summarising such a behaviour of economies of scale exists. Here, a graphical view of the data about scale economies has to precede the stage of fitting the statistical functions to the data.

TECHNICAL FEATURES OF TUBEWELLS

The salient technical information in respect of tubewell is summarised in Table 6.1. As the size of the tubewell, measured in terms of water discharge capacity, increase from 0.014 m^3/second to 0.168 m^3/second, the bore depth of the tubewell rises steadily from about 67 metres to 149 metres, and the length of the slotted pipe – through which water in underground formations surrounding the pipe of the tubewell flows into the tubewell–increases from 1.8 metres to 59.5 metres. These estimates are underpinned, among other things, by the following noteworthy considerations:

(1) Water occurring in first 30 metres (about 100 feet) of boring is not tapped with strainer pipe. It is left untapped with housing pipe.[11]

(2) Fifty per cent of the bore depth in the alluvial formation is water bearing and the rest is non-water-bearing.[12] Thus equal lengths of slotted and blind pipes have been provided in each tubewell.

---

10. Instead of a cubic function, a quadratic form can be employed to represent the theorists' U-shaped cost curve if unit instead of total cost is analysed.
11. The engineering justification for doing so is to minimize the externality of tubewell working on other users of groundwater resources. The success of this measure would depend upon how far the water-bearing strata are inter-connected (hydrologically), *i.e.*, water from one stratum moves to the other below when it is withdrawn from the lower one.
12. This proportion of water-bearing strata is probably based on the information gathered during the boring of numerous tubewells in the Gangetic plains portion of Uttar Pradesh over the last 40 years.

TABLE 6.1

Salient Technical Features of Tubewells

Sr. No.	*Discharge capacity (m^3/second)*	*Bore depth (metres)*	*Bore diameter (cm.)*	*Length of slotted pipe (metres)*	*Pump-set capacity (HP)*	*Length of water course (km.) Lined*	*Length of water course (km.) Unlined*	*Number of delivery tanks*	*Average discharge per water route (m^3/second)*	*Drawdown (metres)*
1.	.014	67	40	19	5.0	0.5	1.0	1	.014	2.8
2.	.028	83	45	26	10.0	1.1	2.1	1	.028	3.8
3.	.042	97	50	33	15.0	1.6	3.2	1	.042	4.6
4.	.056	107	55	39	20.0	2.1	4.3	1	.056	5.2
5.	.070	119	55	45	25.0	2.7	5.3	2	.035	5.7
6.	.084	119	60	45	32.5	3.2	6.4	2	.042	6.5
7.	.098	119	60	49	40.0	3.7	7.5	2	.049	6.9
8.	.112	132	65	51	45.0	4.3	8.5	2	.056	7.4
9.	.126	136	65	53	52.5	4.8	9.6	3	.042	8.0
10.	.140	141	70	56	60.0	5.3	10.7	3	.047	8.4
11.	.154	143	70	57	67.5	5.9	11.7	3	.051	9.0
12.	.168	149	75	60	75.0	6.4	12.8	3	.056	9.3

The standard length of water distribution courses of a common public tubewell of0.042m^3/second discharge capacity is 4.8 km., of which one-third is lined to minimize water seepage to beneficiaries located at a distance from the tubewell. The UPIRI varied this length in proportion to discharge capacity of a tubewell, with the common size tubewell serving as a starting point in estimation. In order to cope with increased tubewell discharge, the water-carrying capacity of the courses is raised suitably, as also the number of routes per tubewell. Thus, the water discharge envisaged per water-course route varies between 0.035 m^3/second and0.056m^3/second for tubewell above the common size. That the envisaged water distribution system for tubewell of high discharge capacity may create serious organizational problem is certainly a matter for concern that cannot be brushed aside in the context of optimum size of a public tubewell. May be, the basis of selling tubewell water would have to be changed from the volumetric basis to area irrigated basis if one has to successfully cope with the task of organizing water supplies to thousands of farmers located far away from the tubewell site.

In the case of augmentation tubewells, the capital and maintenance cost on account of the water distribution system is determined by (a) whether the canal is lined and (b) the extent to which one wishes to minimize the problem of induced seepage from unlined canals. If augmentation tubewells are to pump water into a lined canal, e.g., in Karnal district in Haryana, the distribution cost is negligible as the tubewells can be sited just off the canal 'bunds'. In the case of an unlined canal, tubewells may have to be sited as much as one km. away from the canal, with the result that water channels for conveying tubewell water to the canal have to be constructed. Evidently, the cost in such a case will be much less than that for the water distribution system in the case of a command tubewell.

COST COMPONENTS

The main capital cost items of public tubewell are listed in Table 6.2, while the principal vector of prices underlying these estimates is given in Table 6.3. What is noteworthy about the capital cost structure is the remarkable change in the importance of the four principal segments of a tubewell as the size of a tubewell increases. While electricity transmission segment cost[13] is of greater significance for the lowest size tubewell, water distribution segment comes to figure more and more predominantly as the size of the tubewell increases. This changing capital cost structure is brought out in Table 6.4, in which the percentage shares of the four segments in total capital cost are set out.

Recurring or operating costs of a tubewell are on account of electricity, establishment charges and annual repairs. Within establishment charges one may distinguish between the overhead tubewell organization and the salary of tubewell operator, because both these two items differ for command and augmentation tubewells. In the case of augmentation tubewells, one tubewell operator can mind more than one tubewell simultaneously because he has not to look after the task of water deliveries to different beneficiaries. As regards command tubewells, one tubewell operator per tubewell is the minimum require-

13. Source: Jain (8).

TABLE 6.2
Components of Capital Cost of a Tubewell

Sr.	Tubewell discharge capacity (m^3/second)	Capital cost on account of tubewell assembly						Capital cost of pumping set	Capital cost of water distribution system	Electric transmission	Total capital cost (8+9+(10+11)
		Housing pipe	Slotted pipe	Blind pipe	Boring and lowering of assembly	Miscle-laneous	Total (3+4+5+6+7)				
(1)	(2)	(3)	(4)	(5)	(6)	(7)	(8)	(9)	(10)	(11)	(12)
1.	.014	2,054	1,373	672	6,720	6,822	17,641	12,540	17,004	25,000	72185
2.	.028	4,300	2,751	1,342	8,280	8,350	25,023	14,100	33,942	25,000	98,065
3.	.042	4,300	4,077	1,986	9,660	9,812	29,835	16,500	56,608	25,000	127,943
4.	.056	5,005	5,445	2,650	13,425	11,767	38,292	19,500	73,706	25,000	156,498
5.	.070	5,005	7,488	3,987	14,900	12,926	44,306	22.800	93,320	25,000	185,426
6.	.084	5,005	10,593	6,392	14,900	14,379	51,269	26,100	111,417	25,000	213,786
7.	.098	5,005	12,331	7,437	15,925	15,658	56,356	31,950	128,897	25,000	242,203
8.	.112	7,524	14,114	8,509	16,500	18,478	65,125	34,500	147,694	25,000	272,319
9.	.126	7,524	18,926	12,613	16,975	19,799	75,837	38,250	166,257	25,000	305,344
10.	.140	7,524	20,965	13,970	17,675	22,320	82,454	41,560	183,885	25,000	332,899
11.	.154	8,076	22,744	15,237	17,900	23,943	87,900	43,500	202,625	25,000	359,025
12.	.168	8,459	25,184	16,777	18,625	25,509	94,554	44,400	221,142	25,000	385,096

TABLE 6.3
Vector of Prices Underlying Estimates of Capital Cost of Tubewells

Sr. No.	*Name of the item*		*Unit cost/price*
1.	Housing pipe and blind pipe		Rs. 4 per kg.
2.	Slotted pipe		Rs. 4 per kg. plus Rs. 6 for making 100 slots
3.	Boring charges		Rs. 100 per metre for first 100 m. Rs. 125 per metre for next 50 m.
4.	Unlined water courses		Rs. 1,570 per km.
5.	Lined water courses	--0.014 m^3/second discharge capacity	Rs. 14,260 per km.
		--0.028 m^3/second discharge capacity	Rs. 17,570 per km.
		--0.042 m^3/second discharge capacity	Rs. 21,840 per km.
		--0.056 m^3/second discharge capacity	Rs. 22,080 per km.
6.	Land		Rs. 2,500 per hectare
7.	Well development		Rs. 50 per metre
8.	Pump house		Rs. 3,000
9.	Pea gravel		Rs. 110 per cubic metre

Note :- The list is far from complete.

TABLE 6.4
Changing Cost Structure of a Tubewell

Sr. No. size	*Tubewell Capicity (m^3/second)*	*Percentage share is total capital cost*			
		Tubewell assembly	*Pumpset*	*Water courses*	*Electric transmission*
1.	.014	24.4	17.4	23.6	34.6
2.	.028	25.5	14.4	34.6	25.5
3.	.042	23.3	12.9	44.2	19.6
4.	.056	24.5	12.5	47.1	15.9
5.	.070	23.9	12.3	50.3	13.5
6.	.084	24.0	12.2	52.1	11.7
7.	.098	23.3	13.2	53.2	10.3
8.	.112	23.9	12.7	54.2	9.2
9.	.126	24.8	12.5	54.4	8.3
10.	.140	24.8	12.5	55.2	7.5
11.	.154	24.5	12.1	56.4	7.0
12.	.168	24.6	11.5	57.4	6.5

ment. Also, each tubewell operator needs to be provided with assistance if he has to deal with large number of farmers because of the large size of a tubewell. In arriving at the number of assistants per tubewell operator, the author has been guided by the number of separate delivery tanks per tubewell, whence one assistant for each additional delivery tank is provided. The number of assistants is varied as follows :

Size of tubewell (m^3/second)	*No. of assistants*
0.014 to 0.056	Nil
0.070 to 0.112	One
0.126 to 0.168	Two

Annual cost on account of the salary of a tubewell operator is being provided at Rs. 4,000, and that for the assistant Rs. 2,500. The recurring expenditure on account of overhead organization, which essentially comprises repair and maintenance persons like mechanics and electricians, is being taken at Rs. 1,000 per tubewell, irrespective of the size of the tubewell. Expenditure on account of repairs is being provided for at the rate of 3 per cent of capital cost of a tubewell, even though repair costs may rise less than proportionately with tubewell size.

Cost on account of electricity consumption is determined by (i) horse power of tubewell motor, (ii) annual running hours and (iii) the basis of power tariff. In case of unavoidable power tariff, provision is being made at the rate of Rs. 100 per horse power of pumpset motor.[14] However, electricity cost on an avoidable power tariff is reckoned at Re. 0.15 per kWh.[15]

Besides the operating cost, account has to be taken of (i) interest on capital expenditure and (ii) depreciation charge. In this paper,

14. It may be pointed out that the current unavoidable power tariff for pumpsets in Uttar Pradesh is about Rs. 180 per horse power. Since capital cost on account of transmission segment is borne by the public tubewell, there is need for providing for cost on account of power generation and its transmission on high tension lines only.
15. In the UPIRI memorandum, electricity cost is computed on the basis of Re. 0.15 per kWh. As compared to this price, economic price of electricity for the rural areas is generally considered to be Re. 0.30 to Re. 0.40 per kWh, while the resource cost is taken at Re. 0.25 per kWh.

interest charge is being provided at 10 per cent, which is the rate being charged by the institutional sources financing public investments in rural development. (The basis of depreciation charge has already been spelt out earlier in the methodology section.)

Thus, the annual cost of tubewell irrigation in this paper comprises the following five items :

(1) Electricity bill,
(2) Establishment charges
(3) Expenditure on annual repairs,
(4) Interest charge, and
(5) Depreciation charge.

While the estimates of electricity cost at various levels of working of a tubewell are set out in Table 6.5, the other four components of annual cost are shown in Table 6.6. The estimates of unit cost of tubewell water for various sizes of tubewell are given in Table 6.7.

TABLE 6.5
Electricity Bill

Sr. No.	*Tubewell size (m^3/second)*	*Electricity bill for running hours equal to*						*Electricity bill on basis of flat (unavoidable) tariff*
		2,000	*3,000*	*4,000*	*5,000*	*6,000*	*7,000*	
1.	.014	1,119	1,679	2,238	2,686	3,358	3,918	900
2.	.028	2,338	3,357	4,676	5,371	6,714	7,833	1,800
3.	.042	3,357	5,036	6,714	8,058	10,072	11,751	2,700
4.	.056	4,476	6,714	8,952	10,742	13,428	15,666	3,600
5.	.070	5,595	8,393	11,190	13,429	16,786	19,584	4,500
6.	.084	7,273	10,910	14,546	17,456	21,820	25,457	5,850
7.	.098	8,952	13,428	17,904	21,485	26,856	31,332	7,200
8.	.112	10,071	15,107	20,142	24,171	30,214	35,250	8,100
9.	.126	11,749	17,624	23,498	28,198	35,248	41,123	9,450
10.	.140	13,428	20,142	26,856	32,227	40,284	46,998	10,800
11.	.154	15,107	22,660	30,214	36,256	45,320	52,873	12,150
12.	.168	16,785	25,178	33,570	40,285	50,356	58,749	13,500

TABLE 6.6
Components of Annual Cost (Other than Electricity) of a Tubewell

(Rs.)

Sr. No.	*Tubewell capacity*	*Estblish-ment*	*Repairs*	*Depercia-tion*	*Interest*	*Total annual cost (other than (electricity) (3+4+5+6)*
(1)	*(2)*	*(3)*	*(4)*	*(5)*	*(6)*	*(7)*
1.	.014	5,000	2,166	2,976	7,219	17,361
2.	.028	5,000	2,942	3,840	9,807	21,589
3.	.042	5,000	3,838	4,774	12,794	26,406
4.	.056	5,000	4,695	5,839	15,650	31,184
5.	.070	7,500	5,563	6,861	18,543	38,467
6.	.084	7,500	6,414	7,901	21,379	43,194
7.	.098	7,500	7,266	9,091	24,220	48,077
8.	.112	7,500	8,170	10,160	27,232	53,062
9.	.126	10,000	9,160	11,442	30,534	61,136
10.	.140	10,000	9,987	12,457	33,290	65,734
11.	.154	10,000	10,771	13,298	35,903	69,972
12.	.168	10,000	11,553	14,091	38,510	74,154

TABLE 6.7
Unit Cost of Tubewell Water : Rupees per Thousand Cubic Metres

Sr. No.	*Tube-well size (m^3/ second)*	*On the basis of avoidable power tariff when running hours are*				*On the basis of unavoidable power tariff when running hours are*			
		2,000	*3,000*	*4,000*	*5,000*	*2,000*	*3,000*	*4,000*	*5,000*
1.	.014	183.3	125.9	97.2	79.6	181.2	120.8	90.6	72.5
2.	.028	118.7	82.4	65.3	53.5	116.0	77.3	58.0	46.4
3.	.042	98.4	69.3	54.8	45.6	96.3	64.2	48.1	38.5
4.	.056	88.4	62.7	49.8	41.6	86.3	57.5	43.1	34.5
5.	.070	87.4	62.0	49.3	41.2	85.3	56.8	42.6	34.1
6.	.084	83.4	59.6	47.7	40.1	81.1	54.1	40.5	32.4
7.	.098	80.8	58.1	46.8	39.4	78.3	52.2	39.2	31.3

contd...

8.	.112	78.3	56.4	45.4	38.3	77.1	51.4	38.5	30.8
9.	.126	80.3	57.9	46.6	39.4	77.8	51.9	38.9	31.1
10.	.140	78.5	56.8	45.9	38.9	75.9	50.6	38.0	30.4
11.	.154	76.7	55.7	45.2	38.3	74.1	49.4	37.0	29.6
12.	.168	75.2	54.7	44.5	37.8	72.5	48.3	36.2	29.0

SCALE ECONOMIES IN CAPITAL COST

There are considerable scale economies in capital cost of tubewell irrigation. The source of greater economy is the electricity transmission segment. In terms of our formulation, the scale economy is (+) 1.0 in the size range of tubewells considered.[16] Substantial scale economies exist in the case of pumpset cost as well as main tubewell assembly cost. More precisely, the scale economies are of the magnitude of (+) 0.42 in the case of pumpset cost and (+) 0.30 in the case of tubewell assembly cost. The sources of economy/diseconomy in respect of major components of tubewell assembly are as follows :

Housing pipes	(+) 0.50
Boring and lowering of pipes	(+) 0.57
Blind pipe	(+) 0.62
Slotted (strainer) pipe	(-) 0.21
Miscellaneous components	(+) 0.45

It is in the water distribution segment that one comes across scale diseconomies. Though the magnitude of the coefficient of scale diseconomy appears to be very small, namely,(-) 0.03, the water distribution segment acquires increasing importance with the size of a tubewell. Whether such diseconomies would be experienced in the case of piped water distribution system has not been investigated.[17]

A word about the nature of large economy in pumpset cost would be in order. As per expectations, the size of the horse power of the tubewell

16. There is an element of over-statement here, because our assumption of constancy of installation cost for electricity transmission over the entire size range is too strong to be true in practice. The size of the conductor, the transformer, the pole, etc., is bound to vary in a step manner as the size of the electric motor rises. However, the rise in cost on this is expected to be far less than proprtionate compared to the rise in the size of a tubewell.
17. In view of the very small size of holdings in the densely populated eastern Gangetic plains, the task of land acquisition for the water distribution system becomes diffcult. This has led, among other things, to the use of underground pipes for water conveyance in this region.

motor rises somewhat faster than the rise in the discharge capacity of the tubewell. This would give rise to scale diseconomy if the price of a pumpset were proportional to horse power of the pumpset. Since the manufacturers of pumpsets realise substantial scale economies in the production and marketing of pumpsets, they pass on the benefits to the buyers by following what may be described as an inelastic price policy, i.e., the larger the pumpset, the lower the cost per unit of size of the pumpset.[18]

All said and done, there is evidence of the operation of scale economies in capital cost of a tubewell. Excluding the electricity transmission segment, the scale economy works out to be (+) 0.16, but including the electricity transmission segment the scale economy may be of the order of (+) 0.20.

SCALE ECONOMIES IN OPERATING COST

The three principal operating costs of a tubewell are : energy or electricity bill, repairs to tubewell system and establishment charges. Electricity bill shows a positive evidence of scale diseconomy, whose magnitude is (-) 0.10. The source of this diseconomy is the increased depth from which tubewell water has to be lifted as the discharge capacity of a tubewell increases.[19] It may be mentioned that the magnitude of this scale diseconomy is invariant or insensitive to (a) price of electricity or the basis of power tariff and (b) number of annual running hours for a tubewell.

While the scale economies in the cost of repairs are considerable, one has to keep in mind two points in this connection. One, the cost estimates are open to criticism because these are reckoned simply as 3 per cent of the capital cost. Since there are definite scale economies in capital cost of a tubewell, scale economies in repair costs follow, irrespective of what percentage of capital cost is assumed for the

18. It is partly this inelastic price basis that has promoted the installation of pumpsets/ motors /engines of excessive capacity by farmers in this country. The disbursers of institutional finance are said to have aggravated this phenomenon of excessive capacity, because they may collude with the sellers of pumpsets to sell larger pumpsets so as to inflate the size of the loan to the farmers-on which they are said to obtain illegal cuts from the sellers.
19. In technical terms, the 'drawdown' of a tubewell rises with size, as can be ascertained from Table 6.1.

repair costs. Two, a tubewell with larger command may reduce the timeliness with which repairs in the water distribution system are carried out. As the distance between the tubewell site (where the tubewell operator sits) and the point of fault in the water distribution system increases, time lag in the detection of fault rises. Of course, expenditure of additional cost on patrolling of the water distribution system can solve the problem of timely information.

In short, there are scale diseconomies in operating costs, the magnitude of the scale diseconomy varying with the basis of power tariff and the assumption with regard to the annual running hours per tubewell.

OVERALL SCALE ECONOMIES

The scale diseconomies in operating cost of tubewell irrigation are not strong enough to offset the scale economies in capital cost at any size of tubewell. In point of fact, one does not come across a U-shape in unit cost of tubewell irrigation in the range of tubewell sizes analysed for scale economies in this paper. This result is at variance with the UPIRI result because the methods of analyses and the concepts of costs do differ though the data base is the same in both the studies.[20] More precisely, the scale economies exist upto the size level of 2.5 cusec (0.070 m^3/second) tubewell in the UPIRI study, whereas these prevail upto 6.0 cusec (0.168 m^3/second) tubewell in our paper. However, the two following observations about the behaviour of unit cost of tubewell irrigation need to be noted.

Firstly, in both the UPIRI and our study the tubewell of 2.5 cusec is such that the rise/decline in the unit cost of tubewell irrigation beyond this capacity is of marginal significance. The differences in unit cost, which become narrower with an increase in the number of running hours per year, are small enough to warrant the safe conclusion of constant unit cost beyond 2.5 cusec size. Secondly, the unit cost of tubewell irrigation diminishes rapidly upto the tubewell size of 1.5 cusec (0.042 m^3/second).

20. The result is also at variance with Bhatia and Mehta's finding that a tubewell of 1.0 cusec is of optimum size. Though they consider tubewells in the range 0.34-2.0 cusec, they are primarily concerned with exploring scale along with technological options for shallow tubewells. For more details, see Bhatia and Mehta (1).

POLICY IMPLICATIONS

In the case of public tubewell with independent command area, the Tubewell Irrigation Department in Uttar Pradesh can go in for setting up tubewells of capacity greater than 1.5 cusec, say of 2.5 cusec, as recommended by the UPIRI in its memorandum. While this author, unlike the UPIRI, does not perceive any scale diseconomies beyond this size, he would caution against the setting up of tubewells of capacity greater than 2.5 cusec on two grounds. Firstly, one has to contend with the danger of external diseconomies . Secondly, one must reckon with increasing proportion of evaporation and seepage losses as the size of a tubewell increases. In other words, the effective output of a tubewell, measured by the volume of water finally delivered to beneficiaries at their field inlets, tends to differ more and more from the rated output as the length of the water distribution network increases with tubewell size.

If a public tubewell is to be employed in the role of an augmentation tubewell, the larger the size of the tubewell is, the lower will be the unit cost of tubewell water that is lifted for augmenting canal supplies. However, while exploiting these scale economies, which are internal to the public tubewell, one has to reckon with the external diseconomies of a tubewell. The greater the number of other users of groundwater in the vicinity of a tubewell, the more the external diseconomies. Big tubewells need not cause serious externalities where users of groundwater sources are few. This condition is likely to be fulfilled where water supplies from an augmented public canal improve substantially enough for beneficiaries in the vicinity of the augmentation tubewells so that they no longer need to maintain their own sources of groundwater irrigation. Evidently, this condition cannot be fulfilled if the water raised by the augmented tubewells is earmarked for use at places away from the location of tubewells in whose vicinity many private users of groundwater exist. This is precisely what has created problems in the case of augmentation tubewells of the Western Yamuna Canal in Haryana. Here, the tubewell water is clearly earmarked for Hissar district, whereas the augmented tubewells are located in Karnal district having many private shallow tubewells and other wells.

To conclude, the optimum size for public tubewells is governed inter alia both by internal scale economies and external diseconomies of tubewell irrigation.

REFERENCES

(1) Ramesh Bhatia and Meera Mehta, "Tubewell Irrigation : Analysis of Some Technical Alternatives", *Economic and Political Weekly,* Vol. X, No. 52 (Review of Agriculture), December 27, 1975.

(2) A.S. Chawla *et al.* : Optimum Capacity of a Tubewell, U.P. Irrigation Research Institute, Technical Memorandum No. 46-RR (G-1), Roorkee, 1975 (typescript).

(3) B.D. Dhawan, "Demand for Irrigation : A Case Study of Government Tubewells in Uttar Pradesh", *Indian Journal of Agricultural Economics,* Vol. XXVIII, No. 2, April-June 1973, pp. 59-67.

(4) B.D. Dhawan, "Externalities of New Groundwater Technology on Small Farmers", *Indian Journal of Agricultural Economics,* Vol. XXX, No. 3, July-September 1975

(5) B.D. Dhawan, "Groundwater Exploitation in India : A Study of Internal and External Economics", *Indian Journal of Power and River Valley Development,* Vol. XXVII, Nos. 6 & 7, June-July 1977, pp. 181-196.

(6) B.D. Dhawan, "Tubewell Irrigation in the Gangetic Plains", *Economic and Political Weekly,* Vol. XII, No. 39 (Review of Agriculture), September 24, 1977.

(7) B.D. Dhawan : Groundwater Irrigation in East U.P., Institute of Economic Growth, Delhi September, 1977 (mimeo.).

(8) J.K. Jain : Public Tubewells, Ministry of Agriculture and Irrigation, New Delhi, January 1977 (mimeo.).

(9) C.H. Hanumantha Rao, "Growth of Irrigation in India : An Outline of Performance and Prospects" in Seminar on Role of Irrigation in the Development of India's Agriculture, Seminar Series-XIII Indian Society of Agricultural Economics, Bombay, and the Institute for Social and Economic Change, Bangalore, 1976.

7

Economics of Groundwater Utilisation

Traditional versus Modern Techniques*

There has occurred in India a rapid diffusion of modern groundwater technology: the number of power pumps in operation rose from just 85,000 in 1951 to over 2.75 million in 1971[1] –an increase of a little over 30 times in 20 years. In the same 20 years period, there was hardly any net addition to the number of traditional water-lifts.[2] According to one reliable assessment, modern water-lifts raised about 38 million acre-feet of groundwater in 1971, as against 27 million acre-feet lifted by traditional water-lifts.[3]

It is widely recognised that the State has played a vital role in the introduction of modern water-lifts in Indian agriculture. Though all the open wells equipped with power pumps (also simply referred to as 'pumpsets' in this paper in order to distinguish them from tubewells) are invariably privately owned (and sometimes co-operatively owned), the State has provided a tremendous boost to their growth through (i) massive programmes of rural electrification and (ii) provision of liberal capital finance through institutional sources[4]. In the case of tubewells, however, it decided to develop tubewells under its

---

* This paper was prepared for a groundwater seminar planned by the Department of Groundwater, Rajasthan Government, for 1974 but was not held. It was published in *Economic and Political Weekly* (*Review of Agriculture*), June 1975.

1. Source : Jain, [7], p 13.
2. Source : Reference [8], p 15.
3. Source : Jain, [4] Annexure X.The volumetric unit acre-foot = 325,851 gallons = 1,233.5 cubic metres.
4. The Rural Electrification Corporation of India alone disbursed about Rs. 570 crores during the Fourth Five-Year Plan period (1969-74) for rural electrification schemes meant primarily for energising power pumps for irrigation. About Rs. 650 crores is the share of institutional agencies for supporting farmers' own investment of the order of about Rs. 450 crores for minor irrigation works, including tubewells and pumpsets. *Source* : Reference [5], p 106.

own aegis too; whence number of public tubewells rose from about 3,000 in 1951 to about 20,000 in 1974. During the same period (1951-74), the number of private tubewells shot up from 21 thousand to about 800 thousand.[5]

Considerable research work has been conducted to emphasise the economic superiority of modern water-lift technology over the traditional technology. Such literature certainly helps due appreciation of the rapid growth of modern water lifts – 19 percent per annum for pumpsets and 17 per cent per annum for tubewells (both are compound rates of growth). However, there has not been much systematic analysis of the factors underlying the economic superiority of the new over the old water-lifts. There has been too much preoccupation with highlighting this superiority strictly from the *individual or private viewpoint*. The substantial external diseconomies that are encountered in the exploitation of scarce 'ill-defined ownership' resources [6] – of which groundwater is probably an ideal example– have not received due and careful attention these deserve.[7] The hydrologists' concern, that a decline in the water table in the wake of the tubewell explosion is symptomatic of an undesirable trend towards 'mining' [8] of groundwater resource of the nation, is well taken to the extent that is focuses attention on higher pumping costs with the passage of time, or even outright exhaustion of groundwater in areas

---

5. The figure for 1974 is a tentative estimate of the author. Its basis is data about (i) number of tubewells existing at the end of March 1971 and (ii) estimated annual additions, given in Jain [8], p 17.
6. In India, unlike in the United States, groundwater found underneath a given farm area is not a proprietary resource of the farm owner - in the sense that he can not claim damages from his neighbouring farmers when the latter deplete his water supply through groundwater withdrawals by means of groundwater structures installed well within the boundaries of their own farms. Yet, he alone is legally entitled to instale groundwater tapping appliances on his farm. For externalities arising in the case of ill-defined-ownership resources, see Meade [10], p 34.
7. In a way, the hydrologists' and hydrogeologists' great concern about recent decline in water table in parts of the country experiencing tubewell 'explosion' is a reminder to the nation of long-term externalities of new groundwater technology which the present generation is imposing either on itself (in the future date) or on posterity, or both.
8. Mining of groundwater resource is deemed to occur when the quantum of water withdrawal over the entire rain cycle, of good and bad years, exceeds the groundwater recharge over the rain cycle period. Leaving aside the special case of mining in water-logged areas, mining a resource whose substitutes are unknown is undesirable from the viewpoint of posterity as well.

of poor rainfall or low permeability.[9] That a declining water table could also throw out of farming a large number of small and marginal farmers – who can neither use traditional techniques when the water table recedes to uneconomic depth, nor be able to use 'lumpy' new technology so as to pump water at an economic price–has not received adequate notice in the literature. Misgivings expressed by some scholars in this regard rest straightway on the well known Marxist paradigm of capital displacing labour once capitalist farming begins in the farm sector. The paradigm, in the particular case of pumpset and tubewell capital, has not carried much conviction with agricultural economists, most of them believing that the land-augmenting effect of irrigation (especially if used in conjunction with new bio-chemical farm technology) may generate far greater demand for labour than the labour-displacing effect of modern mechanical method of irrigation.[10]

SCOPE OF THE PAPER

The objective of this paper is to reappraise, both from private and social angles, the modern groundwater technology *vis-a-vis* the traditional one.

The paper is divided into two sections. Salient technical and economic features of both traditional and the modern techniques are analysed in Section I. This section provides the necessary backdrop to our discussion of the external diseconomies of the new groundwater technology in Section II of the paper. We analyse how external diseconomies in groundwater utilisation get accentuated by unregulated introduction of tubewell technology in the Indian agricultural sector. How small and marginal farmers face dire threat to their very

9. Even if apparent exhaustion of groundwater resource may not occur, lowering of water table may lead to infiltration of saline or brackish water from neighbouring areas, thereby rendering groundwater resource unfit for irrigation. Such a danger is to be kept in mind while exploiting groundwater resource in coastal and arid tracts of Western India.
10. Lately, some discerning and farsighted scholars have drawn attention to a new danger, whereby the demand for agricultural labour may increase so steeply that farmers may find the use of labour-saving machinery (e.g., harvest combines) extremely profitable from private viewpoint. In such a situation the intensification in demand for labour due to tubewell-cum-tractor-cum-biochemical technology invites the use of labour-saving machinery, giving rise to potential displacement of labour from the farm sector that cannot be gainfully employed in non-farm sector either.

existence as peasant farmers, is brought out by analysing externalities in a dynamic context.

I

DEPTH OF WATER TABLE

Before we begin with the specific features of various techniques of raising groundwater, it is necessary to highlight the crucial role of the depth of water table (D) in the economics of groundwater utilisation. The operational or variable cost of lifting one unit of water (v) with a given technique is an increasing function of water depth, such that $\frac{dv}{dD} > 0$ and $\frac{d^2v}{d^2D} \geq 0$. Likewise, the unit capital cost (k) ot lifting water is an increasing function of depth of water table, i.e., $\frac{dk}{dD} > 0$ and $\frac{d^2k}{d^2D} \geq 0$. If D* is defined to be the depth of a groundwater structure (e.g., the length of a tubewell bore) $dk/dD^* > 0$ and $d^2k/d^2D^* \geq 0$. To illustrate this point, we have estimated the capital cost of a 6-inch tubewell built in the early sixties in Comilla District of Bangladesh varied with depth of bore as follows :

$$C = 16{,}764 + 0.106\ D^{2*}$$

$$R^2 = 0.664,\ n = 28$$

$$C = \text{Total capital cost.}$$

Evidently, a change in the depth of water table has significant economic implications for an irrigator. For example, a rise in the water table would lower the marginal operational cost of lifting groundwater, whereas a decline in the water table may not only raise marginal operational cost but would also give rise to a situation of diminished water availability – resulting in loss of farm output of the irrigator. Furthermore, a decline in the water table may necessitate additional capital expenditure so as to make the water-lift functional. For instance, both the power pump and its position may require change in the case of permanent lowering of water table.

TABLE 7.1

Labour Requirements of Traditional Water-Lifts

Type of Water-Lift	*Daily Labour Requirements* Human (no. of men)	Animal (no. of bullocks)	Total Labour (no of men equivalents)	*Daily Water Output (gallons)*	*Labour Input Per 10,000 Gallons of Water*	*Average Lift (feet)*
1	*2*	*3*	*4*	*5*	*6*	*7*
I Human-Operated Lifts						
(1) Counterpoise	2		2	5,000	4.0	12
(2) Picotta	5		5	15,000	3.3	20
(3) Chain pump	4		4	18,000	2.2	15
II Animal-Operated Lifts						
(4) Rope-and-bucket	2	2	6	18,000	3.3	30
(5) Persian wheel	2	2	6	22,500	2.2	25
(6) Average for						
(i) human-operated lifts			3.6	13,000	2.9	15.6
(ii) animal-operated lifts			6.6	20,000	3.0	27.5

Notes : (1) In aggregating human and animal labour one bullock-hour is considered equal to 2 human-hours.

(2) Daily work means 10 hours of work during which water is actually being lifted. Thus daily water output is obtained by multiplying hourly discharge rate by factor of ten.

(3) Labour requirements are assessed in terms of persons/animals employed throughout the day, and not by persons/animals actually working *at a time*.

Source : Jain [7], pp 12-13

Fluctuations in the depth of water table occur almost continually under complex interplay of a number of factors, of which the following four are noteworthy :

1. amount of rainfall (R);
2. volume of groundwater withdrawn (W);
3. rate of recharge of ground aquifer from nearby canals, surface reservoirs, and any subterranean groundwater stream (I),
4. rate of groundwater outflow into stream and drains (O).

Of central interest here is the positive relation between groundwater withdrawals and the depth of water table – i.e., for any given area dw/dt and dD/dt are *ceteris paribus* positively correlated. (For a given area dD/dt is negatively associated with dR/dt and dI/dt but has positive association with dO/dt.

TRADITIONAL TECHNOLOGY

A variety of groundwater lifting devices come under the ambit of traditional technology whose principal hallmark is heavy reliance on human or animal labour, both in the construction and operation of such devices. These devices may be distinguished among themselves in terms of factor proportion of human to animal labour in their operational cost. Counter-poise lifts (locally known as *'dhenkali'*, *'lot'*, *'tera'*, *'shadoof'*, etc.), chain pumps, and picotta, are human labour using only, whereas rope-and-bucket lifts (locally known as *'mhote'*, *'charsa'*, *'kapila'*, and *'pun'*) and persian wheels (known as *'rahat'*) use both human and animal labour.

In Table 7.1 is presented a broad comparative picture of the labour requirements of different traditional lifts. On the whole, animal-operated lifts and human-operated lifts appear to have equal labour requirements (about two men-equivalents per 10,000 gallons of water output). But this can be misleading, unless the intensity of labour involved -- as reflected by the height by which water is lifted - is also kept in the reckoning. It can be seen from Table 7.1 that the average lift involved in the case of animal-operated appliances is about 28 feet, as compared to about 16 feet for purely human-operated devices. (Also, labour exertion for human beings is considerably less when men work with animals who bear the brunt of water-lifting.) Thus animal-operated water-lifts are really less labour-using than purely human-operated water-lifts. And the persian wheel may be deemed to be the least traditional technique in the continuum of the traditional technology of lifting groundwater.[11] But, then, it is

11. One is tempted to consider persian wheel as an intermediate technology in the context of the current debate on choice of technology for labour-abundant developing economies. However, such a view may be tenable only under very favourable circumstances of groundwater occurrence, i.e., in regions where (i) copious groundwater supply is available at a low depth of water table within the reach of a persian wheel; (ii) groundwater recharge is extremely good, and (iii) water so available is enough for irrigation needs of the area under cultivation.

comparatively more capital-using among the traditional water appliances.

Since for large farmers[12] family labour is insufficient in relation to their farm needs, the labour cost incurred by them in lifting groundwater is essentially at full market price. Such farmers would find animal-driven devices more economical than purely human-powered devices. Only in the well known case of the 'bonded labour' system – as exists in parts of the Eastern Gangetic plains – can large farmers manoeuvre to get labour services at a price far below the one existing for free labour. In such peculiar circumstances, where paid wage is even below subsistence wage level, farmers may find animal-operated devices not that attractive.

For a small farmer with more than enough supply of family labour (in relation to his small farm size), the labour cost incurred in groundwater lifting may be much below the market price when alternative avocations for surplus family labour are lacking in an economy, Human-powered devices may preponderate in regional economies characterised by (a) small holdings, (b) paucity of avenues for labour absorption, (c) acute prevalence of 'bonded labour' evil, and (d) usurious interest rates on limited investible capital. In economies where all conditions, except the condition (a), are absent (e g, the Punjab and Haryana economies), the question of choice of technology for groundwater may be settled by small farmers over the entire range of techniques – both modern and traditional.

There were over 4 million open wells, fitted with traditional water lifting devices, in operation at the end of the sixties (see Table 7.2). This number is probably not very different from the one existing two decades earlier at the beginning of the planning era. It is not as if no fresh traditional wells were set up by farmers during 1950-70. In fact, over two million open wells were added under various promotional schemes of the state governments. However, many of these were fitted with modern lift devices right in the beginning, and many owners of

12. The dichotomy of farmers, into large and small, requires a word of clarification. The cut-off point is where family labour supply matches farm labour requirements. The supply of family labour resource in the case of large farmers is substantially short of farm labour requirements. The contrary position holds in the case of small farmers.

traditional groundwater structures managed to effect a switchover to modern methods of lifting groundwater.

TABLE 7.2

Statewise Distribution of Traditional and Modern Water-Lifts in 1971

State	*Number of Traditional Lifts*	*Number of Modern Lifts*		
		Pump-sets	*Private Tubewells*	*State Tubewells*
1	*2*	*3*	*4*	*5*
Andhra Pradesh	313,849	165,499	15,738	11
Assam			15	22
Bihar	145,350	79,204	10,239	1,274
Gujarat	385,489	145,673	447	828
Haryana	17,574	46,372	34,259	935
J & K	20			
Madhya Pradesh	517,370	57,123	181	73
Maharashtra	409,543	305,993		5
Karnataka	145,951	125,844	338	
Punjab	76,197	92,675	62,018	845
Rajasthan	599,622	33,439	473	52
Tamil Nadu	497,165	468,215	25,543	23
Uttar Pradesh	978,142	144,418	119,899	9,404
West Bengal	NA	NA	9,458	1,298
All India	4,086,272	1,664,455	278,608	14,829

Source : Reference [4], Annex X.

The current regional dispersal of traditional groundwater structures shows that these are concentrated in Uttar Pradesh, Tamil Nadu, Rajasthan, Maharashtra, Madhya Pradesh, Haryana, Andhra Pradesh and Bihar. To what extent the traditional technology still holds sway in the different states of India is best indicated by the water-lifting capacity of traditional groundwater *vis-a-vis* modern ones. This is a better measure than simply comparing the numbers of traditional and modern structures as it takes care of the following points:

(1) Traditional groundwater structures yield less water compared to modern structures.
(2) Tubewells yield more water than pumpsets.
(3) State tubewells yield more water than private tubewells (see Table 7.3).

At the all-India level, about two-fifths of the groundwater was raised by traditional water-lifts in 1971. Today, this fraction may be still smaller. State-wise variation of the percentage share of traditional lifts in the overall quantum of groundwater lifted was as follows:[13]

Haryana	5
Punjab	12
Tamil Nadu	29
Bihar	33
Maharashtra	36
Karnataka	32
Andhra Pradesh	40
Uttar Pradesh	43
Gujarat	50
Madhya Pradesh	78
Rajasthan	87
All India	42

Finally, it is important to remember the feasible upper limits of the depth of water table (D) with which the various traditional water-lifts can cope. Counterpoise-lifts and chain-pumps become practically unusable when the depth of water table is more than 15 feet. The comparable limiting value for animal-operated water-lifts may be taken to be twice this limit, that is, 30 feet. In the case of the picotta the critical limit can be placed at 25 feet.[14] It is assumed here that these critical limits of 'D' for various appliances are also the economically feasible limits in tapping groundwater by various traditional techniques.

13. These percentages are based on data given in reference [4], Annexure X.

14. These critical depths are nothing but the technical upper limits for which different types of traditional devices can operate. See, Jain, [7], pp 10-11. It need hardly be added that these limits can be extended in the case of a farmer having access to powerful, well built draught animals.

MODERN TECHNOLOGY

Modern groundwater technology embraces two distinct techniques. (1) an open or dug-well equipped with a power pump (being simply referred to as 'pumpset' in this paper) and (2) a tubewell operated with a power pump. (One may as well mention here the 'hybrid' modern technique of a dug-cum-bore well which has characteristic features of both a tubewell and a pumpset.) The hallmark of the modern technique is that the power pump becomes more than a substitute for labour powered water-lift – in the sense that it speeds up the water-lifting process. For instance, a power pump empties collected water of a dugwell in barely two hours, whereas a traditional lift would have taken 10 hours. (This speed-up also results in enhanced water yield from an open well.)

The tubewell technique, however, does much more than this simple substitution of a traditional water-lift by a speedier, energy-based water-lift. It not only alters the groundwater structure (whereby a bore of a few inches cross-section replaces a 'hole' of a few feet cross-section), but it also offers a new concept in tapping underground water-bearing strata: water is not first collected in a man-made storage space underneath and then lifted to the ground surface. Hence tubewell technique may be considered the more modern of the two modern techniques, and our concern is mainly with this technique whenever we allude to modern groundwater technology.

Three noteworthy economic features of the modern technology in comparison with traditional technology are: (1) The unit variable cost of modern technology is extremely low when compared with the corresponding unit variable cost of traditional technology. (2) Capital outlay per modern water-lift is much more than the corresponding outlay on a traditional water-lift. (3) The input needed for creating and maintaining modern groundwater lifts, unlike in the case of traditional lifts, have to be principally procured with cash payment from a number of sources in the non-farm sector of the economy.

LOW OPERATIONAL COST

It is well recognised that, a little inanimate energy (steam, electricity and petroleum oil) can replace a lot of animate energy being expended by man and beast in any mechanical activity. Depending upon the depth of water table, one to two kilowatt hours (kWhs) of inanimate energy can displace both the drudgery and the energy expended by a man or a beast in a full day's work in lifting water with traditional lifts. The market daily wage rates for human and bullock labour are known to be far higher than the price of one kWh of inanimate energy, with the result that the variable cost of a modern groundwater lift becomes remarkably low in comparison with the variable cost of a traditional lift.[15] (To a small extent, the state policy of supplying electricity to the farm sector at a subsidised rate has, indeed, widened the cost differential in question.) By and large, the advantage in respect of variable cost arises from the fundamental fact that the resource cost of producing a unit of inanimate energy is substantially lower than that of producing a comparable unit of animate energy.

HIGH CAPITAL NEEDS

Installing a power pump on an open well entails a sizeable capital outlay, which is much greater than the outlay on a comparable traditional water-lift . The difference between the two capital costs becomes smaller as one moves from purely human water-lifts to the persian wheel. In fact, the difference appears rather marginal between a pumpset and a persian wheel. But this is so strictly from a farmer's individual or private viewpoint. The situation changes radically as one views capital costs from a national angle. To be concrete, it is well known that the nation has to incur substantial capital expenditure (with a good deal of foreign exchange component) to create the necessary infrastructure in the form of electricity generation, transmission, and distribution system which makes available motive power for electric pumpsets and electric tubewells. This infrastructural investment – not to mention capital investment in industries making metal tubes, pumpsets, etc. – is far higher than what an individual farmer spends on his own account while installing a power pump. Just to give an order of the dimensions involved, if the cost of an electric pumpset was Rs. 1,500[16] in the mid-sixties, the

15. For sake of illustrating the dimension of costs involved, Table A sets out the comparative values of 'v'.
16. This estimate is taken from Shukla, [17], p 49.

corresponding infrastructural investment of the nation in the electricity supply system was probably about Rs. 7,500 per electric pumpset connection.[17]

TABLE 7.3

Statewise Unit Estimates of Groundwater Withdrawal for Various Types of Groundwater Structures

(Acre feet)

S No	*Name of the State*	*Traditional Open Well*	*Pumpset*	*Private Tubewell*	*State Tubewell*
1.	Andhra Pradesh	5	12	20	200
2.	Bihar	8	16	40	300
3.	Gujarat	8	16	40	200
4.	Haryana	8	20	40	250
5.	Madhya Pradesh	5	12	40	250
6.	Maharashtra	5	12	-	200
7.	Karnataka	5	12	20	-
8.	Punjab	8	20	40	250
9.	Rajasthan	8	20	40	300
10.	Tamil Nadu	5	12	15	200
11.	Uttar Pradesh	8	20	40	300

Source : Reference [5], Annex X.

In case the nation is unable to come forward with investment in the electricity infrastructure, an individual farmer would have to marshall another Rs. 3,000 for the installation of an oil engine to provided drive to his power pump.[18] Viewed in this light, the much-applauded programmes of rural electrification have indeed sharply reduced the

17. In computing a comparable infrastructural investment for an electric pump we have assumed (a) the investment cost of electricity generation capacity is Rs. 1,000 per horse power, (b) the transmission and distribution cost averages at Rs. 4,000 per pumpset connection, and (c) the horse power of the electric motor of the pumpset is 3.5 HP. It may be mentioned that the component (b) is exclusive of cost on account of high tension (HT) lines that bring power from the generating point to low tension (LT) lines of the rural areas.
18. The source of this figure is the same as in footnote 16.

capital needs of the farmers for installing new groundwater devices. Coupled with the fact that the operational cost of an electric tubewell/pumpset is significantly below the operational cost of a diesel tubewell/pumpset, one can well appreciate why the rate of growth of diesel-driven power pumps has lagged behind the rate of growth of the electrically-driven power pumps – 15 per cent compared to 26 per cent per annum during 1951 to 1971. Notwithstanding the fact that the proportion of diesel power pumps has steadily declined, from three-fourths in 1951 to just about two-fifths in 1971, one may still be puzzled by the rather high annual increment to the stock of diesel pumps in the country. This brings us to the speed of rural electrification, the sharp outward shift in demand for irrigation water over which a farmer has control in timely application, and the disbursement of institutional funds for financing capital outlays on tubewells and pumpsets. We shall discuss first the last one which ties up with our present exposition of capital needs of the new groundwater technology.

In farm sectors, characterised by low productivity per person, the process of capital accumulation is a painful one – whence investible capital (other than the potential capital that can be created through appropriate programmes designed to convert surplus human labour into productive assets) becomes an extremely scarce resource. The scarcity reflects itself in intolerably high interest rates, which permit investment either in truly money-spinning assets or divert investible funds to distress consumption needs for which usurious interest rates can be extracted from the victims of natural calamities. In such a situation, mainly medium and large farmers in canal-irrigated tracts are capable of generating enough internal surpluses to pay the cost of a tubewell or a pumpset. If they perceive that canal water supply is inadequate, or erratic, or totally unavailable at certain crucial periods of sowing or maturing of a crop, they might invest in the new technology on their own. Since old as well as new canal works have one or more of these deficiencies (which have been heightened by the new bio-chemical farm technology which makes rather exacting demands on irrigation water), even medium and small farmers in canal tracts have made substantial investments of their own in the new groundwater technology – especially when they succeed in getting part of the cake of liberal institutional finance.

In the absence of rural electrification, even the medium and large farmers in non-canal tracts could not take rapid options on the new groundwater technology without liberal capital assistance from outside. And access to this outside institutional capital happens to be vitally tied up with the size of the owned holding (or political influence which again bears some association with farm size). Thus, large farmers have been ideally placed for taking full advantage of institutional capital finance for new groundwater technology. Wherever they have been hesitant to invest in new technology, despite access to outside finance capital, one major inhibiting factor among others has been fragmentation of their farm holding.

SIZE AND FRAGMENTATION DISABILITIES

Pumpset/tubewell being a lumpy investment with a minimum floor value, farm acreage which is to benefit from such investment acquires great importance in the adoption of modern groundwater technology. To the extent the floor value is lowered as a result of technological advance in industries making power pumps, electric motors, oil engines, and tubewell construction equipment, the minimum farm acreage for an economically viable investment in modern technology also becomes smaller. However, one should not be too optimistic on this account. Quite often, prices of motors, pipes, etc., are highly inelastic with respect to the size variable. So any technological advance in making equipment of smaller sizes does not get appropriately reflected in lowered prices for equipment of lower denomination or rated capacity. Since the variable costs of modern technology are already rather low, any economies effected in operational cost, via the technological advance *a la miniaturisation* style, may be inconsequential in reducing the minimum necessary farm acreage for which investment in modern groundwater technology is an economic proposition. This is probably one of the main reasons why farmers and the promoters of modern technology find it not worth their while to do very fine tuning with regard to the selection of the size of the motor for pumps. Undoubtedly, there is considerable social or national loss involved, however, in the shape of more than requisite consumption of scarce power – as tubewell and pumpset owners employ electric motors of capacity greater than technically necessary.[19]

19. The author's ideas in this area are developed elsewhere. See Dhawan, [3].

TABLE 7.4

Radius of Influence and Area of Influence of a Tubewell

	Type of Underground Soil	*Radius of Influence (feet)*	*Area of Influence (acres)*
1.	Fine sand layers with some silt	100-300	1-7
2.	Fine to medium sand layers, fairly clean and free from silt and clay.	300-600	7-26
3.	Coarse sand and fine gravel layers free from silt and clay	600-1000	26-72
4.	Coarse sand and gravel without silt and clay	1000-2000	72-289

Source : Reference [12], p 10

Large farmers may still suffer from overcapitalisation of their farms in an indirect manner. This stems from lack of consolidation of their holdings, and to distinguish this kind of size disability from the earlier 'pure' one emanating from size of the farm, this is being labelled here as 'fragmentation disability'. Only a fraction of large farmers can be free from the fragmentation disability of the new technology, wherever the size of single biggest parcel of their land happens to be equal to or above the land size mark below which the apprehension of size disability exists.[20] Wherever the state governments implemented land consolidation reforms, they have provided an additional impetus to the diffusion of new groundwater technology. The penetration of the new technology is likely to be hampered in regions where small or/and fragmented farms predominate in the structure of land holdings.

Now we may as well discuss the indirect methods by which both the size and fragmentation disabilities of modern technology can be surmounted. One way open to a farmer is to think in terms of either setting up co-operative pumpsets and tubewells or entering into arrangements to sell and buy water from each other. While precise

20. Apparent fragmentation disability of an unconsolidated farm may disappear if the owner can manage to interconnect some parcels of his land via water courses to such an extent that the farm area so 'consolidated' indirectly exceeds the minimum farm size free from size disability.

information about the extent of co-operatively-owned pumpsets and tubewells is lacking, the general impression is that such wells have not made much headway in India. Instead, many tubewell and pumpset owners have managed to effect substantial sales of water to their neighbouring farmers. On the basis of the limited number of surveys available so far it appears that the extent of such sales of water is associated with the natural co-operative spirit pervading in a region – as indicated by the success of co-operative movement in that region. For example, such sales are reported to take place on a greater scale in Gujarat than in the states in the North-West region of India.

Another way in which a small farmer (or a large farmer having a fragmented holding) can overcome the size/fragmentation disability of the new technology is to use the motor power of the oil engine/electric motor for purposes other than water-lifting. This route is becoming popular in the north-western Indian states. For example, tubewell owners in Punjab, Haryana and UP, according to a PEO study, utilised their motors for almost one-third of their annual running time for non-irrigation purposes – such as fodder-cutting, or oil-miling, or dal miling, etc.[21] It is not known, however, how far small farmers and large farmers, with fragmented holdings – as distinguished from large farmers, with well-consolidated holdings - have overcome their size or fragmentation disability in this manner.

The third way to surmount both the size and the fragmentation disabilities is for the state to establish state-owned tubewells. However, state tubewells do not appear to be an economic proposition in consolidated poor water-bearing strata of the Indian peninsula (1.2 million square km) from which enough yield of water (with the possible exception of trap formations) cannot be obtained to spread thinly the incidence of a number of unavoidable overheads that are specific to state tubewells only – e.g., (i) salary of the tubewell operator, (ii) supporting organisation for tubewell repair and maintenance, (iii) lodging for tubewell operators, (iv) electricity transmission from HT to LT (whenever such costs are not borne by the state electricity boards), etc. Where the water-bearing strata are unconsolidated but shallow (e.g., coastal Andhra Pradesh), the water yield may be enough to operate shallow tubewells. Unless such tubewells are operated in clusters – one cluster under one tubewell operator – these may again

21. See Reference [6].

prove uneconomic if operated under state auspices.[22] In short, hydro-geological conditions would largely determine in which regions the state tubewells can help small farmers and large farmers with fragmented holdings, in case they are unable to do so on their own initiative. Thus the Gangetic plain which covers Central UP, Eastern UP, and Northern Bihar, appears an ideal region for the development of public tubewells. To the extent the irremediable flood proneness of this area is an additional disincentive to private investments in fixed assets, the state should socialise the risk by undertaking investments on its own.[23] The real hurdle in these areas in successfully managing the public tubewells would be the problem of widespread graft and the problem of public property becoming easy prey to so called 'anti-social' elements.

METHODS OF FINANCIAL APPRAISAL

Assumptions underlying the financial criteria followed for providing institutional finance for agricultural development have been questioned by scholars – notably economists. Economists have been unhappy with the criterion because of its deviation from the standard benefit-cost criterion. The standard benefit-cost criterion views a project not only in terms of its contribution in a macro-context but also corrects, via shadow pricing, for the project input-output prices which are believed to arise from imperfections in the markets of such inputs and products. Some projects that are fit for selection on benefit-cost criterion may not be viable if the financial criterion is applied to them (and *vice versa*). Notwithstanding the fact that economists have failed to indicate how the financial institutions should resolve the dilemma in such awkward situations, irrigation projects like private tubewells and pumpsets, really do not pose this kind of dilemma.

---

22. One has also to reckon with the possibility of inflow of sea water into ground aquifers adjacent to sea shores in the event of heavy withdrawals of groundwater by heavy-duty state tubewells.

23. I consider the flood-proneness of this area irremediable for two reasons. First, the size of the population in this area has increased to such a level that substantial part of the population has now settled into 'khadar' or riverine lands. Even medium floods can play havoc with the fixed assets of such a population. Secondly, control of floods in valleys adjacent to hill ranges is a difficult task when a good deal of the hill areas with high seasonal rainfall happen to lie inside the boundaries of another nation (in this case Nepal).

The rub of the matter is two-fold : (a) the repayment capability of the beneficiary gets largely determined by the size of his own farm and its degree of fragmentation, and (b) the time-horizon fixed for repayment of loan is less than the life of the asset being considered for finance. Thus, even if a farmer does not suffer from disability pertaining to either size or fragmentation, the factor (b) could disqualify him from getting financial assistance from an institution. The real effect of the procedure underlying factor (b) is to raise the minimum economic farm size for which the new groundwater technology is viable, and the procedure underlying the factor (a) does not take into account the fact that farmers do sell water to their neighbours. The scope for good revenue income from this 'agro-business' is totally neglected by the institutional financiers. Since the marginal cost of water sold to others is far below the average cost[25] (as well as the price which buyers are willing to pay),[26] small farmers as well as large farmers inhabiting tracts known for 'co-operative spirit' (an euphemism for simply perceiving mutual advantage in a transaction) ought to receive fair consideration from financial institutions on this account.

Even if the tenanted holdings are free from size and fragmentation disabilities, the tenants cultivating these holdings cannot take advantage of institutional sources of finance without the initiative and concurrence of the owners of such lands. This, in turn, requires that tenants and landlords must work out mutually satisfactory relations for appropriate sharing of irrigation costs as well as for consequently increased agricultural output. Such prospects are rather dim. Thus, regions having a good deal of either tenant-cultivated holdings or lands with disputed titles (e.g., in the eastern Gangetic plains) are ill-placed to take full advantage of institutional finance for tapping groundwater with capital-intensive groundwater techniques, till appropriate institutional reforms occur.

---

24. In this regard see reference [2].
25. To illustrate, the reported average and marginal costs of pumping groundwater in Gujarat state are as in Table B.
26. Some illustrative statistics of price realised from sale of water in Meerut district are as in Table C.

TABLE 7.5

Extent of Lowering of Water Table at Various Distances for a Tubewell.

Tubewell No	*Observed decline in water table (in feet) at a distance (in feet) of*							*Remarks*
	2	50	100	250	500	1000	2000	
I.	7.0	6.5	5.8	4.5	3.3	2.0	1.0	This tubewell of 3.8 cusec discharge rate was operated for 30 days
II.	3.2	2.5	1.8	1.2	0.7	0.3	--	This tubewell of 2.9 cusec discharge was operated for 10 days

Source : Reference [1].

II

EXTERNAL DISECONOMIES

The ramifications of externalities, arising while tapping groundwater resources, are being increasingly appreciated as the complex nature of this resource is being understood by hydrologists and hydro-geologists. For analytical purposes, we may distinguish the following types of external diseconomies:

(a) A farmer withdraws so much water from an aquifer that he either raises the marginal cost of raising groundwater for other farmers, or lowers the marginal farm product of groundwater of other farmers.

(b) A farmer withdraws so much water from an aquifer in year 't', that he raises his own (or of his heirs) marginal cost of lifting groundwater in subsequent years t+1, t+2,....

(c) A group of farmers, tapping groundwater in a given area, withdraw so much water that it either reduces the marginal cost of lifting groundwater elsewhere or lowers the marginal farm product of groundwater elsewhere.

(d) A group of farmers withdraw so much volume of water that it leads to a reduction in 'base' flow of a nearby effluent river, thereby affecting adversely the interests of beneficiaries of river water.

In this paper we are not really concerned with externality (d), except for recording two observations. First, this externality underscores the need for a systems approach to water resource utilisation. To illustrate with a concrete example : the 'base' flow in the Yamuna river has been significantly reduced as a result of excessive development of tubewells in Haryana and West UP, from whence originates a substantial groundwater outflow in the 'effluent' Yamuna river. This in turn has reduced the irrigation capacity of the Agra Canal taking off from Okhla (near Delhi).[27] Secondly, the optimum economic depth upto which tubewell technology should tap groundwater resource will depend both on the internal economies of tubewell operation as well as by the spillover effect of a lowered water table on the river flow. To bring home the point, while it may be economic for a tubewell to be-operated at a depth of water table D, at this depth the externality (d) may be very powerful (so much so that an 'effluent' river might turn 'influent').

LONG-TERM EXTERNALITIES

The hydrologists/hydro-geologists have indeed done signal service to the nation by alerting policy makers to grave long-run consequences of excessive groundwater withdrawal with rapid but unregulated introduction of tubewell technology. Their warning with regard to groundwater 'mining' in certain tracts[28] experiencing rapid diffusion

27. It is said that the irrigation capacity of Agra Canal has been reduced by 50,000 acres. But since this canal could hardly give more than two 'rabi' irrigations per farmer, the 'output effect' of diminished water availability would be tremendous in the case of farmers in the tail end of canal. A team of the Central Groundwater Board is investigating the causes of diminished water flow in the Yamuna river, and would determine if Haryana's heavy duty (4 to 6 cusecs) canal augmentation tubewells dug along the river area have been a contributory factor. It is said that these deep tubewells have also seriously affected the working of private tubewells in their vicinity.

28. In Gujarat state, Mehsana district and Ahmedabad city have experienced an alarming decrease in the depth of water table. In parts of Mehsana district the tubewell explosion has lowered the water table from 40-50 feet 15 years ago to 200 feet in 1974. Moreover, the depth of boring for a tubewell has risen from 300 feet to 1,000 feet if the water output is to remain intact. Likewise, "static water levels (in Ahmedabad city) have gone down at a rate of 6 to 10 feet per year. Static water levels, which were at depths of 40 to 50 feet in 1939, have reached 165 to 170 feet in certain parts of the city" (*Source*: Patel and Shah, [13], pp 4-85). Excluding groundwater withdrawn by cotton mills and other mills in Ahmedabad, the Ahmedabad Municipal Corporation alone raised its groundwater withdrawal from 1.70 billion gallons in 1951-52 to 18.35 billion gallons in 1971-72. (*Source*: Ramanathan, Patel and Shah, [16]).

of the tubewell technology have not, however, been appropriately heeded in the state legislatures.[29] One important economic explanation, as to why these warnings have failed to register themselves on the legislators as well as the users of the new groundwater technology, is the extremely low value of unit operational cost (v) of the new technology compared to the corresponding value for traditional groundwater technology. Consequently, lowering of the water table does not give rise to serious cost implications for a farmer owning a modern water-lift. For him, the real cause for concern arises when the lowered water table diminishes his water yield. This eventuality, too, has been averted by the adopters of new groundwater technology, by and large, by measures designed to maintain intact their groundwater output.

Because of a number of relieving circumstances, the pinch of the much-talked about declining water table is not being acutely felt. First, the sharp outward shift in the demand curve for irrigation water has helped to cushion the additional expenditure entailed by the decline in water table. Secondly, the lowering of water table in canal-irrigated tracts has come as a welcome relief to policy makers working under the shadow of water-logging. Thirdly, massive extension of the rural electrification network has not only reduced a farmer's capital need for powering his tubewell/pumpset but has also reduced his unit operational cost wherever he has managed to effect a switchover from a more costly oil-driven to a less expensive electric-power pump. Fourthly, continued expansion in institutional finance for farm equipment, on liberal terms, has softened the pinch of extra capital outlay which would otherwise have to be raised from the village capital market at exorbitant interest rate.

It is true that, in the matter of access to liberal finance, small and marginal farmers have not been as fortunate as large farmers backed with bureaucratic-cum-political influence. But, then, small farmers have been relative late-comers to the new groundwater technology. It

29. Water being a state subject, the Centre has been urging state governments to enact suitable groundwater legislation to prevent excessive withdrawal of groundwater by tubewells. In this connection, a model bill -- titled "The Groundwater (Control and Regulation) Bill 1970" - was prepared by the Centre to guide the state legislatures. So far, no state has adopted any legislative measure in the matter of groundwater utilisation.

is, after all, small and marginal farmers, having surplus bullock/family labour, of low opportunity cost on the one hand, or/and lack of access to the liberal segment of the capital finance market, on the other hand, who still find using traditional groundwater-lifts a worthwhile, if not paying, proposition. Those among them, who opted for the new groundwater technology, may have been hurt badly by the decline in the water table – especially if they failed to raise the additional finance required to maintain intact the water yield of their wells.

It is primarily the users of high-cost traditional water-lifts who experience sizeable external diseconomies due to permanent lowering of water table by the introduction of new groundwater technology in their vicinity. In case the water table recedes to a level which is still within the *technical capability* of their water-lifts, the externality shows up either as enhanced cost of lifting water or diminished water output, or both. Given that (a) technical capability of human-operated counterpoise water-lifts is almost half the capability of animal-operated water-lifts and (b) average cost curve of animal-operated water-lifts is below that of the other traditional lifts, the magnitude of external diseconomies consequent upon a certain recession in water table would be far greater for owners of counterpoise appliances than for owners of animal-driven appliances. And, should the water table recede 30 feet below ground surface, it threatens to throw out of farming activity the users of traditional modes of groundwater lifting, unless they can effect a changeover to modern techniques of tapping groundwater.

How disastrous may be the consequences of the long-run externalities on the owners of traditional water-lifts would depend upon both the magnitude of decline in water table and the ability of these owners to effect appropriate capital changes in their wells. In the event the water table is lowered just below the technical limits of a human water-life, the scope for a switchover to animal-driven devices may be exploited by deepening the dug-well. In case the water table has receded below the lifting capacity of farm animals, one may not be able to carry on farm activity without changing over to new groundwater technology. In either case, additional capital outlays are needed. It is here that the small and marginal farmers may find themselves stranded. To begin with, their lack of access to the liberal segment of the rural capital market may stand in their way. Secondly, even if they can manage to raise the additional capital required for a switchover in techniques, their size and fragmentation disabilities may be substantial enough to make the additional investment look uneconomically unattractive. Unless they see the prospect of surmounting these disabilities in some indirect manner (say, either by selling surplus water to neighbours, or buying water from neighbours using new groundwater technology), they are doomed as peasant-farmers.

SHORT-TERM EXTERNALITIES

Adverse short-term externalities of the tubewell technology on the users of traditional water-lifts and open wells fitted with power pumps – to be distinguished from short-run externalities which tubewell owners impose on one another due to lack of proper spacing among tubewells - are not so well realised in the literature as are the long-term externalities of the unregulated introduction of the tubewell technology. In order to bring out these externalities it is necessary to dwell here on induced changes in water table brought by pumping in the vicinity of a tubewell. As water is pumped out, water held in the formations surrounding the tubewell begins to infiltrate into the tubewell. The surrounding formations, so affected, are in the shape of a solid cone whose base is on the surface of the static water level and the apex touches the surface of the pumping level of groundwater. The radius of the circular base of this inverted cone is known as the 'radius of influence' (r) of the well, and as pumping level recedes in the well 'r' becomes larger.

The value of 'r' is vitally dependent on two factors: groundwater withdrawal rate dw/dt and the physical characteristic of the water-bearing strata. dw/dt is of high magnitude for modern groundwater technology, as compared to traditional technology – with the result that the area of influence of the former technology is several times that of the latter technology. Thus, compared to 5 to 8 acre-feet of water withdrawn annually through a traditional open well, 12-20 acre-feet is raised by an open well fitted with a power pumpset; 15-40 acre feet by a shallow tubewell; and 200-300 acre-feet by a deep state-owned tubewell (see Table 7.3).

A tubewell installed in soil having good transmissibility of water has a radius of influence which is almost 10 times as much, and an area of influence that is almost 40-100 times that of a tubewell installed in soil with poor transmissibility -- e.g., fine sand with some silt (see Table 7.4). Thus, the scope for the emergence of externalities is much greater among the coarsely sandy water-bearing strata than in the finely sandy ones.

Within the radius of influence of a well the extent of adverse externality – as reflected in the lowering of the water table – experienced by another groundwater exploiter is in inverse proportion to his distance from the well in question. How the water table is lowered in open wells, at various distances from a tubewell, has not been empirically and thoroughly investigated in India. More importantly, we do not have studies indicating the lowering of water table at various distances as a result of the *simultaneous* operation of a few tubewells with radii of influence overlapping. In order to illustrate the potential short-run hydraulic spillover-effects on users of groundwater resources at different distances from the site of a *single tubewell*, we present in Table 7.5 some data based on pumping tests for two tubewells of West Pakistan. It is clear from such data that small and marginal farmers operating traditional water-lifts in an area having tubewells may be experiencing severe external diseconomies when tubewells depress water table in the very short run. Such temporary depressions in the water table, which are likely to occur at peak times of agricultural season when tubewells have a high probability of working simultaneously, may have a deleterious impact on crops under the command of traditional water-lifts. The prospect for such an eventuality is all the greater when the expected rains fail. Thus, the very reliability of tubewell water during droughts can be a source of undoing of the farmers using traditional lifts – unless the former agrees to supply water to their doubly-stricken brethren.

CONCLUDING REMARKS

Given the skewed distribution of land ownership and the small size of the average farm, the adverse externalities of the tubewell technology cannot be completely avoided in a free enterprise framework, even if the state legislatures pass legislation to control and regulate the use of groundwater. However, the externalities can be internalised if a public agency undertakes to supply groundwater. In fact, public tubewells – wherever they are feasible – are not only the best means for managing groundwater resources efficiently, but also the only way to overcome the problem of lack of utilisation of groundwater resource in certain areas afflicted by fragmented, tenanted, small holdings.

TABLE A

State tubewell	Rs 1.4 per cubic 1000 cubic meteres
Private tubewell	Rs 1.2 per cubic 1000 cubic metres
Persian wheel	Rs 44.6 per cubic 1000 cubic metres
	Rs 13.0 per cubic 1000 cubic metres

Source : Based on data in Moorti, [11], pp 61-64.

TABLE B

Type of Pumpset	*Cost Per Acre-Inch of Water*		*Number of Hours Worked in a Year*
	Average	*Marginal*	
Oil-dirven	9.6	4.9	700
Electric	6.0	3.6	920

Source : Based on pumpset-wise data given in Table 5.2 of Patel and Patel, [14]

TABLE C

	Percentage of Lifted Water Sold	*Average Price (Rs per acre-inch)*	*Marginal Cost (Rs per acre-inch)*
Electric tubewell	43	4.84	1.93
Diesel tubewell	24	5.68	3.42
Pumpset	39	7.36	3.83

Source : Patel and Patel, [15], p 48 and p 77.

REFERENCES

1. Ahmed, Nazir, 'Hydraulics of Tubewells'. Engineering News, Volume 12, Number 4, December 1966.
2. Dhawan, B D, 'Utilisation of Groundwater Resources: Public vs Private Tubewells', *Economic and Political Weekly*, Review of Agriculture, September 28, 1974.
3. "A Note on Power Requirements of Tubewells", a paper presented to Conference on Power for Growth organised by the Punjab, Haryana and Delhi Chamber of Commerce and Industry held at Delhi in February, 1975 and published in the "Compendium of Background Papers", published by PHDCCI.
4. Government of India, Planning Commission, Report of the Task Force on Groundwater Resources, Delhi, 1972, (Mimeo).
5. --Draft Fifth Five-Year Plan, Part II, Delhi, 1973.
6. --"Study of Tubewell Irrigation and Groundwater Development Programme", Programme Evaluation Organisation, New Delhi, April 1974, (Mimeo: draft report).

7. Jain, J K, "Groundwater Structures", New Delhi, 1973, (Mimeo).
8. –"Tubewell Construction in India", New Delhi, 1972 (Mimeo).
9. Khan, A Aziz, "Tubewell Irrigation in Comilla Thana", Pakistan Academy for Rural Development, Comilla, Bangladesh, July 1972 (Mimeo; second edition).
10. Meade, James E, "The Theory of Economic Externalities". Institute Universitative de Hautes Etudes Internationales. Geneva, 1973.
11. Moorti, T V "A Comparative Study of Well Irrigation in Aligarh District India": *Cornell International Agricultural Development Bulletin,* Number 19, May 1971.
12. Murty, V V N, K C Jain and R N Pandey, "Tubewells"; Communication Centre, Punjab Agricultural University, Ludhiana, 1973.
13. Patel, A H and S R Shah, 'Groundwater and Water Supply Problems of Ahmedabad', "Proceedings of International Symposium on Development of Groundwater Resources", November 1973.
14. Patel, S M and K V Patel, "Studies on Economics of Rural Electrification and Lift Irrigation", Indian Institute of Management, Ahmedabad, 1969.
15. –, "Economics of Tubewell Irrigation," Indian Institute of Management, Ahmedabad, 1971.
16. Ramanathan, K R, A H Patel and S R Shah, "Suggestions for Augmenting the Yield of Water from the Infiltration Wells System at the Headworks at Dudeshwar, Ahmedabad", A paper presented at the Seventh Annual Convention of Indian Water Works Association, Bombay, held at Madras in January 1975.
17. Shukla, V P, 'Interaction of Technological Change and Irrigation in Determining Farm Resource Use. Jabalpur District, India, 1967-68', *Cornell International Agricultural Bulletin*, Number 20, July 1971

8

Groundwater Exploitation in India

A Study of Internal and External Economics*

INTRODUCTION

Modern groundwater technology, based on power pumps, has rapidly diffused in Indian agriculture, notably during the last ten years. Between 1951 and 1966 the number of power pumps rose from about 83 thousand to 984 thousand, and in 1974 this number crossed the 4 million mark.[1] There has been a veritable tubewell 'explosion' since the mid-sixties. Today, there are almost a million tubewells in existence, compared to just 6 thousand tubewells in 1951.

On the other hand, there has been hardly any net addition to the stock of dugwells (open wells) fitted with manually or animal-operated water-lifting devices. Today, such traditional wells account for no more than one-fourth of the total volume of groundwater being raised annually in the country. In point of fact, these have practically gone out of use in some parts of the country that have witnessed successful introduction of new biochemical technology.

The current rates of annual increment of traditional and modern water-lifts are as follows :[2]

---

* This paper was published in the *Indian Journal of Power and River Valley Development,* Vol. 27, Nos 6 and 7, June-July, 1977.

1. Of the 4.193 million power pumps in operation in 1974, a small fraction (say about 50,000) was being used for surfacewater lift-irrigation.
2. These estimates are based on trends in the latter half of the Fourth Plan period. We are presuming that these trends are continuing unchanged.

	Numbers	*As percentage of stock in 1974*
Dugwells fitted with traditional lifts	nil[3]	0
Pumpset wells (dugwells fitted with power pumps)	250,000	7
Shallow tubewells	100,000	12
Deep tubewells	1,000	5

On the supply side, the State has played a vital role in the introduction of modern water-lifts in Indian agriculture through (a) its massive programmes of rural electrification and (b) provision of substantial capital finance at liberal interest rate through institutional sourcesunder its direct control. According to one reliable estimate for the early seventies, of the Rs. 400 crores being invested annually in groundwater development, farmers' own finance was of the order of Rs.80 crores only.[4] Other sources of finance were as follows :

1. Institutional outlay	Rs. 120 crores
2. Outlay on rural electrification	Rs. 150 crores
3. Public sector outlay[5]	Rs. 50 crores
Total 1 to 3	Rs. 320 crores

THE PROBLEM

Considerable research work has been done, highlighting economic superiority of the modern groundwater technology over the traditional technology. But the factors underlying this superiority have not received a fuller probe. What is more significant to note is that the superiority has been viewed from private viewpoint, and that too from a static or short-run angle. Both the short-run and the long-run 'external effects' of the new groundwater technology have been neglected. The hydrologists and the hydrogeologists alone have repeatedly warned

3. Actually, about 170,000 new dugwells are created. But because of installation of power pumps on old and new dugwells there is hardly any net addition to the stock of traditional dugwells.
4. Source : Reference (4), p. 14.
5. This is actually total public sector outlay on all minor irrigation works. Assuming (i) 1,000 public tubewells were being added annually and (ii) the construction cost (exclusive of electricity transmission) of each public tubewell was Rs. 120,000, the public sector outlay on public tubewells comes to Rs. 12 crores.

about the danger of a decline in water table due to excessive and unregulated exploitation of groundwater resource by means of tubewell technology. They have thus done a signal service to the nation by pointing out how the present generation imposes a cost both on itself (at a future date) and on future generations by lowering water table by excessive pumping of groundwater.

However, external diseconomies of tubewell technology on the users of traditional wells have not been adequately discussed in the literature.

SCOPE OF THE PAPER

The paper is divided into four sections. How depth of water table plays a crucial role in the internal economics of groundwater resource utilisation is discussed analytically in section I. A few empirical relationships of capital cost as a function of depth of tubewell bore are spelt out. Salient techno-economic features of traditional wells, and their place in groundwater exploitation, are discussed in section II. The same is done for modern groundwater technology in section III. Finally, external diseconomies in exploiting a 'common property' resource like groundwater are analysed, both from the short-run and the long-run viewpoints.

One objective of this paper is to bring out how small and marginal farmers face dire threat to their very existence as independent farmers in the wake of rapid tubewell growth in their vicinity.

I
CRUCIAL ROLE OF DEPTH OF WATER TABLE

Before we begin with the specific features of various techniques of raising groundwater it is necessary to highlight the crucial role of depth of water table (D) in the economics of groundwater utilisation. The operational or avoidable cost of lifting one unit of water (v) with a given technique is ceteris paribus an increasing function of water depth, such that $dv/dD > 0$ and $d^2v/d^2D > 0$. Likewise, the capital cost (k) of lifting water is an increasing function of depth of water table i.e. $dk/dD > 0$. If D^* is defined to be depth of a groundwater structure (say, the length of a tubewell bore), $dk/dD^* > 0$ and $d^2k/d^2D^* > 0$.

A change in the depth of water table has significant economic implications for an irrigator. For example, a rise in water table would lower the marginal operational cost of lifting groundwater, whereas a decline in water table may not only raise marginal operational cost but also give rise to a situation of diminished water availability, resulting in loss of farm output. Furthermore, a decline in water table may necessitate additional capital expenditure so as to make the water-lift functional. For instance, both the power pump and its position may require change in the case of permanent lowering of water table. Likewise, increase in D in an area may necessitate increase in D* for a given groundwater structure in order to maintain its water output in the wake of a lowered water table.

Fluctuations in depth of water table occur almost continually under the complex interplay of a number of factors, of which the following four are noteworthy :

1. amount of rainfall (R);
2. volume of groundwater withdrawn (W);
3. rate of recharge of ground aquifer from nearby canals, surface reservoirs and any subterranean groundwater stream (I);
4. rate of sub-surface groundwater outflow into streams and drains (O).

Of central interest here is the positive relation between groundwater withdrawals and the depth of water table *i.e.* for any given area dw/dt and dD/dt are *ceteris paribus* positively correlated.

Empirical information about appliance-wise cost as a function of depth of water table, or depth of groundwater structure, is not available. In what follows some fragments of such information are presented mainly for purposes of illustration. Energy consumption is the major avoidable cost in raising groundwater. In the case of modern groundwater appliances the consumption of inanimate energy is governed by the following linear function of depth of water table.[6]

$$\text{H.P.-hours} = \frac{\text{V.D.}}{273.\text{n}} ;$$

6. *Source* : Zapata, (20), p. 13.

where: H.P. is horse power of pump;
D is depth of water table (meters);
V is the water discharge (cubic meters);
n is pump efficiency (per cent).

In the case of tubewells we have endeavoured to estimate some empirical relationships of capital cost as a function of depth of the tubewell bore. These relationships are derived by us from a cross-section data for 34 diesel tubewells installed in the Comilla area of Bangladesh between 1962 and 1964. Each tubewell is of a uniform 6-inch diameter, with depth of the bore varying considerably between 118 feet and 263 feet (mean value 192 feet).[7]

We tried to fit a simple quadratic regression equation of the type $a + bD^* + cD^{*2}$. Though R^2 of such equations were good, a strong multicollinearity emerged between D^* and D^{*2}, forcing us to drop one of these from the analysis. We have retained the D^{*2} term, because the relationship so emerging still satisfies the basic assumption that $d^2k/d^2D^* > 0$. Also, such relationship showed slightly higher R^2 values than the corresponding simple linear relationships. Below are presented the selected quadratic relationships.

$$K_1 = 3{,}473 + \underset{(18.658)}{0.056\, D^{*2}},\quad R^2 = 0.9158$$

$$K_2 = 881 + \underset{(7.691)}{0.091\, D^{*2}},\quad R^2 = 0.6489$$

$$K = 16{,}764 + \underset{(7.164)}{0.106\, D^{*2}},\quad R^2 = 0.6508$$

Where: K is total capital cost of a tubewell;
K_1 is capital cost on account of tubewell materials other than engine;
K_2 is capital cost on account of boring;
D^* is the depth of the tubewell.
(Figures in brackets are t-values).

One major reason why the coefficient of determination (R^2) is not very high in the case of total capital cost function is that almost half

7. For more details see Khan, (10), Appendix B.

of this cost comprises diesel engine cost which in this cross-section is practically invariant with regard to D*. This invariance has arisen because diesel engine cost is practically the same for all the tubewells under consideration. (This constancy in engine cost is probably due to insignificant variation in depth of water table in the Comilla tract.)

II
GROWTH OF TRADITIONAL WELLS

About 3.5 million traditional dugwells were in existence in 1951. Adding to this progressively the number of fresh dugwells reportedly constructed in subsequent years, the Minor Irrigation (MI) Division of the Central Ministry of Food and Agriculture has computed that the number of dugwells rose to about 5.9 million in 1969, and 6.9 million in 1974. If we deduct from these numbers the corresponding figures of dugwells fitted with power pumps, we can estimate the number of traditional dugwells in existence in these years. Thus, the stock of traditional dugwells was about 4.2 million in 1969 and about 4 million for 1974. If one keeps in mind that the MI Division has made no allowance for dugwells (and other wells) going out of existence during 1951-74, one might conclude that the number of dugwells operated by traditional water lifts in 1974 was probably the same as in 1951 i.e. 3.5 million.

A variety of groundwater lifting devices come under the ambit of traditional technology whose principal hallmark is heavy reliance on human or animal labour, both in the construction and operation of such devices. These devices may be distinguished among themselves in terms of factor proportion of human to animal labour in their operational cost. Counter-poise lifts (locally known as 'Dhenkali', 'Lot', 'Tera', 'Shadoof', etc.) chain pumps and picotta are human labour using only, whereas rope-and-bucket lifts (locally known as 'Mhote', 'Charsa', 'Kapila', and 'Pun') and Persian wheels (known as 'Rahat') use both human and animal labour. But we do not have device-wise break up of 3.5 million traditional water lifts now in existence.

SALIENT TECHNO-ECONOMIC FEATURES OF TRADITIONAL WATER-LIFTS

In Table 7.1 is presented a broad comparative picture of labour requirements of different traditional lifts. In the preparation of this table the following tacit assumptions have been made by this author:

1. In aggregating human and animal labour, one bullock hour is considered equal to two human hours.
2. Daily water output is obtained by multiplying hourly water discharge of a lift by a factor of 10.
3. Labour requirements are assessed in terms of persons/and animals employed throughout the day for getting daily water output specified in assumption (2). Thus number of persons/ animals actually working on a lift at a point of time is not the criterion of labour requirements. To illustrate, while two persons are shown as the labour requirement of a counter-poise lift in Table 7.1, only one person at a time works the lift.

On the whole, animal-operated lifts and human-operated lifts appear to have equal labour requirements, about three men-equivalents per 10,000 gallons of water output. But this can be misleading, unless intensity of labour involved--as reflected in the height by which water is lifted--is also kept in the reckoning. It can be seen from Table 7.1 that the average lift involved in the case of animal-operated appliances is about 28 feet as compared to about 16 feet for purely human-operated devices. (Also, labour exertion for human beings is considerably less when men work with animals who bear the brunt of water lifting). Thus animal-operated water-lifts are really less labour using than purely human operated water-lifts. And the Persian wheel may be deemed to be the least traditional technique in the continuum of traditional technology of lifting groundwater. But, then, it is comparatively more capital-using among the traditional water appliances.

In order to assess the external effects of a permanent or temporary recession in water table it is of crucial significance to know the value of 'D' beyond which a traditional lift is unusable, either for economic or technical (or both) reasons. According to one authority on groundwater,[8] the range of value of 'D' for which different traditional water-lifts are 'suitable' are broadly as follows :

1. Counter--poise lift	6-15 feet
2. Picotta	15-25 feet
3. Chain pump	Under 15 feet

8. Jain, (7), pp. 10-11.

4. Rope-and-bucket lift	25-30 feet
5. Persian wheel	5-30 feet (and even much more)

It is argued here that tapping groundwater by traditional modes tends to become infeasible when the depth of water table is below 30 feet of the surface.

Since for large farmers[9] family labour is insufficient in relation to their farm needs, for them labour cost incurred in lifting groundwater is essentially at full market price. Such farmers would find animal-driven devices more economical than purely human-powered ones. Only in the well-known case of 'bonded labour' system – as exists in parts of the Eastern Gangetic plains – can large farmers manoeuvre to get labour services at a price far below the one existing for free labour. In such peculiar circumstances where the paid wage is even below the level of subsistence wage level, farmers may find animal-operated devices not that attractive.

For a small farmer with more than enough supply of family labour (in relation to his small farm size), labour cost incurred in groundwater lifting may be much below the market price when alternative avocations for surplus family labour are lacking in an economy. Thus, human-powered devices may preponderate in regional economies characterised by (a) small holdings, (b) paucity of avenues for labour absorption, (c) acute prevalence of 'bonded labour' evil, and (d) usurious interest rates on limited investible capital. In economies where all conditions, except the condition (a), are absent (e.g. the Punjab and Haryana economies), the question of choice of technology for groundwater may be settled by small farmers over the entire range of techniques, both modern and traditional.

CONTRIBUTION OF TRADITIONAL LIFTS IN GROUNDWATER EXPLOITATION

According to the Task Force on Groundwater Resources, the annual water output of a traditional dugwell ranges from 5 to 8 acre-feet compared to 12 to 20 acre-feet for a pumpset well, 15 to 40 acre-feet for

9. Categorisation of farmers, into small and large, requires a word of clarification. The cut-off point in our analysis is sufficiency of family labour for farm operations.

shallow tubewell and 200 to 300 acre-feet for a deep tubewell (Table7.2). Thus, a switchover from traditional to modern technology is much more than a simple choice among water lifting devices: it is also a question of more versus less water.

Assuming an average annual output of 6.61 acre-feet per traditional well, the total volume of water raised by 3.5 million traditional water-lifts comes to about 23 million acre-feet for 1974. This works out to a little over 10 per cent of the total groundwater potential of India (215 million acre-feet),[10] and about 23 per cent of the total groundwater raised in 1974. The contribution of traditional water lifts in overall groundwater extraction has substantially declined since 1951, when it stood at about 90 per cent.

Uptodate information about the regional dispersal of groundwater structures is not readily available. The picture of regional towards the end of the sixties shows that these are concentrated in Uttar Pradesh, Tamil Nadu, Rajasthan, Maharashtra, Madhya Pradesh, Haryana, Andhra Pradesh and Bihar (Table 7.2).

State-wise volume of groundwater lifted by different types of water-lifts in 1969 is presented in Table 8.1. On the basis of this table, State-wise variation of the percentage share of traditional lifts in the overall quantum of groundwater lifted was as follows:

1.	Haryana	..	5
2.	Punjab	..	12
3.	Tamil Nadu	..	29
4.	Bihar	..	33
5.	Maharashtra	..	36
6.	Mysore	..	32
7.	Andhra Predesh	..	40
8.	Uttar Pradesh	..	43
9.	Gujarat	..	50
10.	Madhya Pradesh	..	78
11.	Rajasthan	..	87
	All India		42

10. There is some lack of consensus about the groundwater potential of India. The estimate of 215 million acre-feet is the one accepted by the Planning Commission's Task Force on Groundwater. For more details, see reference, (4) pp. 26-27.

TABLE 8.1
Statewise Withdrawal of Groundwater, 1968-69

(Thousand acre-feet)

State	By traditional lifts	By modern lifts			Total groundwater withdrawal
		Pumpset wells	Private tubewells	State tubewells	
Andhra Pradesh	1,569	1,986	315	2	3,872
Assam	-	-	-	4	4
Bihar	1,163	1,267	410	382	3,539
Gujarat	3,084	2,913	18	165	6,180
Haryana	141	927	1,370	234	2,672
Jammu & Kashmir	-	-	-	12	12
Madhya Pradesh	2,587	685	7	18	3,297
Maharashtra	2,048	3,672	-	1	5,721
Karnataka	730	1,510	7	-	2,247
Punjab	610	1,853	2,481	211	5,155
Rajasthan	4,797	669	19	16	5,501
Tamil Nadu	2,486	5,619	383	5	8,493
Uttar Pradesh	7,825	2,888	4,796	2,821	18,330
West Bengal	NA	-	189	260	449
All -India	27,040	24,308	9,995	4,131	65,472

Source : Reference (4), Annexure X.

The regional picture given above is for the year 1969. Since then, the number of pumpset wells has more than doubled, and the number of tubewells has almost tripled. Therefore, the present picture of regional contribution of traditional wells in groundwater exploitation is likely to be radically different from the one given above. For instance, it is said that traditional wells have fallen in total disuse in Punjab and Haryana.

III
MODERN GROUNDWATER TECHNOLOGY

Modern groundwater technology comprises two distinct techniques, tubewells and pumpset well. Whereas a pumpset well is simply a modern version of a traditional well–change of traditional water-lift by a water-lift powered by electricity or oil–, tubewell embodies a new concept in tapping groundwater resource. It does away with the concept of storage of groundwater at the bottom of an open man-made pit, which serves as a receptacle for water infiltrating from soil around the pit. Instead, it taps directly water held by sandy soils, known as aquifers.

Tubewells are classified into shallow and deep tubewells. While the depth of tubewell bore is the criterion for this classification, deep tubewells have generally much greater water discharge rate than shallow tubewells because of greater thickness of aquifers tapped by them. Thus, deep tubewells are capable of irrigating much larger acreage than the shallow one can. In view of size of land holdings and their high fragementation in India, few large farmers in the country are in a position economically to exploit the large water output of a deep tubewell. Thus, deep tubewells have developed under the aegis of State Governments, and have come to be known as State or public tubewells as distinguished from private (shallow) tubewells. The growth of public and private tubewells has been as follows:

	Private tubewells (in '000)	Public tubewells (in '000)
1951	3	3
1961	20	8
1966	113	13
1974	810	20

Speed is a distinct hallmark of modern technology, and this is also true of modern groundwater technology. Thus, modern water-lifts, as embodied in power-driven pumpsets, work faster than traditional water-lifts in raising groundwater. For instance, a power pump can empty collected water of a dugwell in just one-fifth of the time it may take by a traditional water-lift. This speed-up in water lifting results in much more than a mere saving in time – which is of value in itself in agricultural activity where timeliness of operations is of vital importance –, as it also considerably improves water yield of an open well. In short, a switchover from a traditional to a modern water-lift adds significantly to a farmer's ability to bring more of his land under well irrigation, as well as to improve the timeliness of his irrigation operations.

THREE NOTEWORTHY COST FEATURES

Three noteworthy cost features of the modern technology in comparison to traditional technology are: (1) The unit variable cost of modern technology is extremely low when compared to the corresponding unit variable cost of traditional technology. (2) Capital outlay per modern water-lift is much more than the corresponding outlay on a traditional water-lift. Today, a shallow tubewell costs about Rs. 25,000 to farmer, and a deep tubewell costs Rs. 120,000 to a State Government. (These costs are exclusive of supporting infrastructural investment in electricity generation, transmission and distribution). (3) The input needed for creating and maintaining modern groundwater lifts, unlike in the case of traditional lifts, have to be principally procured with cash payment from a number of sources from the non-farm sector of the economy. We shall now discuss in detail features (1) and (2).

Low Operational Cost

It is well recognised that a little inanimate energy (steam, electricity and petroleum oil) can replace much of animate energy expended by man and beast in any mechanical activity. Depending upon the depth of water table, one to two kilowatt hours (kWh) of inanimate energy can displace both the drudgery and the energy expended by a man or a beast in a full day's work in lifting water with traditional lifts. The market daily wage rates for human and bullock labour are known to be far higher than the price of one kWh of inanimate energy, with the result that variable cost of a modern groundwater lift becomes remarkably low in comparison to variable cost of a traditional lift.

The comparative picture of variable costs is aptly brought out by the following statistics of unit variable cost per 1000 cubic metres of water raised:[11]

Public tubewell	..	Rs.1.4
Private tubewell	..	Rs.1.2
Persian wheel	..	Rs. 44.6
Rope-and-bucket lift ('Charsa')	..	Rs.93.0

11. *Source* : Based on data in Moorti, (11), pp. 61-64.

By and large, the advantage of modern groundwater technology in respect to variable cost arises from the fundamental fact that resource cost of producing a unit of inanimate energy is substantially less than that of producing a comparable unit of animate energy.

Private and Social Capital Costs are High

Installing a power pump on an open well indeed entails a sizeable capital outlay, which is much greater than the outlay on a comparable traditional water-lift. The difference between the two capital costs becomes smaller as one moves from purely human water lifts to the Persian wheel. In fact, the difference appears rather marginal between a pumpset and a Persian wheel. But this is so strictly from a farmer's individual or private viewpoint. The situation changes radically as one views capital costs from a national angle. It is well known that the nation has to incur substantial capital expenditure (with a good deal of foreign exchange component) to create necessary infrastructure in the form of electricity generation, transmission and distribution system that makes available motive power for electrically operated tubewells and pumpset wells. This infrastructural investment – not to mention capital investment in industries making metal tubes, pumpset etc. – is far higher than what an individual farmer spends on his own account while installing a power pump. Just to give an order of the dimensions involved, if the cost of an electric pumpset was Rs. 1500[12] in the mid-sixties, the corresponding infrastructural investment of the nation in the electricity supply system was probably of the order of Rs. 7,500 per electric pumpset connection.[13]

In case the nation is unable to come forward with investment in electricity infrastructure, an individual farmer would have to marshall another Rs.3,000[14] for the installation of an oil engine to provide drive to this power pump. Viewed in this light, the programmes of rural

12. This estimte is taken from Shukla, (17), p. 49.
13. In computing infrastructural investment for an electric pump we have assumed(a) the invesment cost of electricity generation capacity is Rs. 1,000 per horse power, (b) the transmission and distribution cost averages at Rs. 4,000 per pumpset connection and (c) the horse power of the electric motor of the pumpset is 3.5 hp. It may be mentioned that the component (b) is exclusive of cost on account of high tension (HT) lines that bring power from the generating point to low tension (LT) lines of the rural areas.
14. The source of this figure is the same as in the footnote 12.

electrification have indeed sharply reduced the capital needs of the farmers for installation of new groundwater devices. Coupled with the fact that operational cost of an electric tubewell/pumpset is significantly below the operational cost of a diesel tubewell/pumpset, one can well appreciate why the rate of growth of diesel-driven power pumps has lagged behind that of the electrically-driven pumps -- 15 per cent compared to 26 per cent per annum during the period 1951-71.[15]

In farm sectors characterized by low productivity per person, the process of capital accumulation is a painful one, whence investible capital (other than the potential capital that can be created through appropriate programmes designed to convert surplus human labour into productive assets) becomes an extremely scarce resource. The scarcity reflects itself in intolerably high interest rates, that permit investment either in truly money spinning assets or divert investible funds to distress consumption needs for which usurious interest rates can be extracted from the victims of socalled natural calamities. In such a situation, mainly medium and large farmers in canal-irrigated tracts are capable of generating enough internal surpluses to pay for the cost of a tubewell or a pumpset. It they perceive that canal water supply is inadequate, or erratic, or totally unavailable at certain crucial periods of sowing or maturing of a crop, they might invest in the new technology on their own. Since old as well as new canal works have one or more of these deficiencies (which have been heightened by the new bio-chemical farm technology that makes rather exacting demand on irrigation water), even medium and small farmers in canal tracts have made substantial investment on their own in the new groundwater technology, especially when they succeed in getting part of the cake of liberal institutional finance.

In the absence of rural electricification, even the medium and large farmers in non-canal tracts could not take quick options on the new groundwater technology without liberal capital assistance from outside. And access to this outside institutional capital happens to be vitally tied

15. The growth of diesel and electric pumps has been as follows:

	1951	1961	1966	1971
Electric pumps (in thousands)	19	193	513	1,629
Diesel pumps (in thousands)	66	230	471	1,150

Source : Jain, (7), p. 13.

up with the size of the owned holding (or political influence which again bears some association with farm size). Thus large farmers have been ideally placed for taking full advantage of institutional capital finance for new groundwater technology.

SURMOUNTING FARM SIZE AND FRAGMENTATION DISABILITIES

Pumpset/tubewell being a lumpy investment with a minimum floor value, farm acreage that is to benefit from such investment acquires great importance in the adoption of modern groundwater technology. To the extent the floor value is lowered as a result of technological advance in industries making power pumps, electric motors, oil engines, and tubewell construction equipment, the minimum farm acreage for an economically viable investment in the modern technology also becomes smaller. However, one should not be too optimistic on this account. Quite often, prices of motors, pipes, etc. are highly inelastic with respect to the size variable. So any technological advance in making equipment of smaller sizes is not getting appropriately reflected in lowered prices for equipment of lower denomination or rated capacity. Since the variable costs of the modern technology are already rather low, any economies effected in operational cost, via the technological advance *à la miniaturisation* style, may be of little consequence in reducing the minimum necessary farm acreage for which investment in modern groundwater technology is an economic proposition. This is probably one of the main reason why farmers and the promoters of the modern technology find it not worth their while to do very fine tuning with regard to the selection of the size of motors for their pumps. Undoubtedly, there is, however, considerable social or national loss involved in he shape of more than requisite consumption of scarce power if tubewell and pumpset owners employ electric motors of capacity greater than technically necessary.[16]

Large farmers, who have farm size above the floor value necessary for economic use of new groundwater technology, may suffer from overcapitalisation of their farms in an indirect manner. This stems from lack of consolidation of their holdings, and to distinguish this kind of size disability from the earlier 'pure' one emanating from size of the

16. The author's ideas in this area are developed elsewhere, See, Dhawan, (3).

farm, this is being labelled here as 'fragmentation disability'. Only a fraction of large farmers can be free from fragmentation disability of the new technology, wherever the size of single biggest parcel of their land happens to be equal to or above the land size mark below which the apprehension of size disability exists.[17] Whenever the State Governments implemented land consolidation reforms, they have provided an additional impetus to the diffusion of new groundwater technology. The penetration of the new technology is likely to be hampered in regions where small or/and fragmented farms predominate in the structure of land holdings.

We may now as well discuss indirect methods by which both size and fragmentation disabilities of the modern technology can be surmounted. One way open to farmers is to think in terms of either setting up cooperative pumpsets and tubewells or entering into arrangements to sell and buy water from each other. While precise information about the extent of cooperatively-owned pumpsets and tubewells is lacking, the general impression is that such wells have not made any headway in India. Instead, many tubewell and pumpset owners have managed to effect substantial sales of water to their neighbour farmers. On the basis of limited number of surveys available so far it appears that the extent of such sales of water is associated with the natural cooperative spirit pervading in a region--as indicated by the success of cooperative movement in that region. For example, such sales are reported to take place on a greater scale in Gujrat than in States in the north-west region of India.

Another way in which a small farmer (or a large farmer having a fragmented holding) can overcome the size/framentation disability of the new technology is to use the motor power of oil engine/electric motor for purposes other than water lifting. This route is becoming popular in the north-west Indian States. For example, tubewell owners in Punjab, Haryana and U.P., according to a PEO study, utilized their motors for almost one-third of their annual running time for non-irrigation purposes such as fodder cutting, or oil milling, or dal milling, etc. it is not however known how far small farmers and large farmers

17. Apparent fragmentation disability of an unconsolidated farm may disappear if the owner can manage to inter-connect some parcels of his land via water courses to such an extent that the farm area so 'consolidated' indirectly exceeds the minimum farm size free from size disability.

with fragmented holdings – as distinguished from large farmers, with well-consolidated holdings–have overcome their size or fragmentation disability in this manner.

The third way of surmounting both the size and the fragmentation disabilities is for the State to establish State-owned tubewells. However, State tubewells do not appear to be an economic proposition in consolidated poor water-bearing strata of peninsular India (1.2 million square km) from which enough yield of water (with the possible exception of trap formation) cannot be obtained to spread thinly the incidence of a number of unavoidable overheads that are specific to State tubewells only e.g. (i) salary of the the tubewell operator, (ii) supporting organization for tubewell repair and maintenance, (iii) lodging for tubewell operators, (iv) electricity transmission from HT to LT (wherever such costs are not borne by the State Electricity Boards), etc. Where the water-bearing strata are unconslidated but shallow (e.g. coastal Andhra Pradesh), the water yield may be enough for operating shallow tubewells. Unless such tubewells are operated in clusters – one cluster under one tubewell operator – these may again prove uneconomic if operated under State auspices.[18] In short, hydrogelogical conditions would largely determine in which regions the State tubewells can help small farmers and large farmers with fragmented holdings, in case they are unable to do so on their own initiative. Thus the Gangetic plain that covers central U.P., eastern U.P., and northern Bihar appears an ideal region for the development of public tubewells. To the extent the irremediable flood proneness of this area is an additional disincentive to private investments in fixed assets, the State should socialize the risk by undertaking investments on its own. The real hurdle in these areas in successfully managing the public tubewells would be the problem of widespread graft and the problem of public property becoming easy prey to so called anti-social elements.

EXPLOITATION OF WATER RESOURCES BY TUBEWELL TECHNOLOGY

Volume of groundwater extracted by tubewell technology rose from about 1.22 million acre-feet in 1951 to 2.95 million acre-feet in 1961,

18. One has also to reckon with the danger of intrusion of sea water into ground aquifers adjacent to sea shores in the event of heavy withdrawals of groundwater by heavy-duty public tubewells. This can damage groundwater resource for ever.

14.13 million acre-feet in 1969, and about 34.63 million acre-feet in 1974. Thus about 16 per cent of the groundwater potential is now exploited with the help of tubewells. Together with pumpset wells, the modern groundwater technology helps in tapping now about 29 per cent of the groundwater resource potential of the country, and about 75 per cent of the actual groundwater output of the country. Thus, new groundwater technology has come to pay a major role in the utilisation of groundwater resources of the country.

At the regional level, uptodate information is lacking. Of the 540 thousand private tubewells existing in 1971, U.P., Punjab and Haryana accounted for 425 thousand i.e. almost fourth-fifth of the total.[19] Likewise, these three States accounted for over 70 per cent of the 16,788 public tubewells existing in 1971.

IV
RAMIFICATION OF EXTERNAL DISECONOMIES

Groundwater is an ideal example of a resource whose ownership is ill-defined : a farmer alone is legally entitled to put up a groundwater tapping device on his farm but he is not the exclusive owner of groundwater underneath his farm. So, for all practical purposes groundwater becomes a common property resource whose exploitation by one farmer may impose external diseconomies on his neighbours who also are utilising the same source of water.

Groundwater is a rather complex resource. The ramifications of external effects can be far reaching, sometimes spilling over much beyond the confines of a given groundwater-bearing tract. In case the rate of exploitation is really excessive in relation to rate of replenishment (recharge) of the aquifer /aquifers in the tract, the externality may extend to users of streams and rivers in the vicinity of the tract. And these users of river water may well be located quite far away from the tract. For analytical insights, the external effects may be divided into the following four types :

(a) Self-imposed externalities;
(b) Intra-aquifer externalities;

19. Source ; Jain, (9), p. 17.

(c) Inter-aquifer externalities;
(d) Trans-aquifer externalities;

Self-Imposed Externalities

To bring out how this paradoxical externality arises, let us assume there is only one farmer tapping an aquifer. In case his rate of groundwater withdrawal tends to exceed the rate of groundwater recharge of the aquifer – a real possibility in poorly recharged aquifers–, the water table may be lowered permanently. This would raise *ceteris paribus* his own (and that of his heirs') marginal cost of extracting water in perpetuity. Should the decline in water table be such that (a) no alteration is required in his water-lift for operating it at newer depth and (b) the groundwater structure still continues to yield requisite volume of water, the externality may show up in the shape of increment to variable cost bill (provided he does not behave as a profit maximizer who equates marginal cost with marginal benefit from groundwater.) But if he acts as a profit maximizer his intake from the aquifer may be diminished somewhat, with the result that his farm output diminishes accordingly. The elasticities of marginal variable cost curve for water and marginal benefit curve for water would determine the extent of diminution in farm output and the increment to variable cost bill.

In case the decline in water table is steep, the farmer may be forced to change over to a water-lift of greater capacity. Generally, a decline in water table leads to a significant decline in water yield from a well. To avoid consequent sizable farm output effect the farmer may have to undertake sizable investment in structural modifications of his well, such as deepening of well, converting an open well into a dug-cum-bore well, lengthening the bore or strainer pipes of a tubewell, etc. (or, he may invest in measures for preventing losses of water in conveyance, or evapo-transpiration). In some situations, the lowering of water table can also give rise to another kind of powerful output effect via reduction in quality of groundwater. In areas where saline or brackish groundwater occurs below or adjacent to fresh water aquifers (e.g. near sea coast and arid areas), quality of groundwater of a fresh-water aquifer may be damaged due to intrusion of saline or brackish water. Consequently, farm output may suffer a set-back in the wake of a decline in water table. (Since farmers in India rarely get the quality of their extracted groundwater tested regularly, the cause of setback to farm output due to the use of saline or brackish water may not be discerned soon enough after the permanent lowering of water table.)

Intra-Aquifer Externalities

This is the common type of externality one hears about in connection with groundwater resource use. In this case, the external effects of one farmer's actions extend to other beneficiaries tapping the same aquifer. That is, as a result of his effort to tap groundwater either the marginal cost of extracting groundwater of some other beneficiaries rises, or there is reduction in their share of availability of groundwater, or both.

For the occurrence of this kind of externality it is not necessary that water table of an aquifer is lowered permanently. For, it can arise even without a permanent recession in water table. In point of fact, a single farmer's actions are not likely to lower water table permanently unless a set of strong conditions is satisfied. Some of these conditions are : (i) rate of replenishment of an aquifer is very poor, (ii) water production of his groundwater tapping appliance is substantial in relation to the combined capacity of other beneficiaries and (iii) the combined water extracting capacity of all the beneficiaries is in excess of the rate of recharge of the aquifer.

It is in the short-run that a single farmer, using a swift mode of extracting groundwater, may impose perceptible external diseconomies on his neighbours, even though there is no permanent decline in water table. Such external effects show up in the shape of reduced water yield at certain times of crop season, with consequent adverse farm output effect. These are technically attributed to mutual interference (over-lapping of 'areas of influence') among wells arising from lack of proper spacing among wells. It is this aspect of mutuality which makes the external effects of groundwater resource use really mutual.

Inter-Aquifer Externalities

Wherever aquifers are well-connected underneath, that is, water from one flows into the other under influence of gravity, excessive exploitation of one aquifer by its beneficiaries may adversely affect the interests of some or all beneficiaries in neghbouring aquifers. (dD/dt is inversely related to dI/dt). Such adverse developments may, however, appear with some time lag after intra-aquifer externalities have become very powerful.

Trans-Aquifer Externalities

Externalities, when they extend beyond the users of groundwater resource to users of river water, may be described as 'trans-aquifer externalities'. These would normally arise when the water table in an aquifer is lowered permanently by a sizable level. For, dD/dt is positively associated with dO/dt (groundwater outflow to an 'effluent' river[20]). If the 'base' flow of a river is committed for some useful purpose (surface irrigation, drinking, navigation, fisheries or waste water disposal),reduction in river flow would adversely affect this use.

The trans-aquifer externalities of groundwater utilisation are now being increasingly appreciated. Though these are not of direct concern to us in this paper, it is still instructive to comprehend their significance. Therefore, we present below a concrete case of such externalities.

This is the case of Yamuna river. In the catchment area of this river that falls within Haryana and U.P. there has been a phenomenal tubewell expansion, so much so that some areas have been witnessing continuous lowering of water table. At the same time there has been a perceptible decline in the 'base' flow (flow not accounted by rainfall run-off in catchment area) of this river, giving rise to reduced availability of water for feeding the Agra canal that takes off from the river at Okhla near Delhi. This canal irrigated about 4 lakh acres (kharif plus rabi) of land in 1963-64 season (as per Administration Report of the Irrigation Department, U.P. for 1963-64), and according to reliable circles its irrigation potential has diminished by 50,000 acres due to diminished base flow of the Yamuna.[21]

20 A river is said to be 'effluent' when it receives groundwater sub-surface outflow. Most rivers are normally effluent, except at flood time when they may turn 'influent' i.e. river water seeps into adjacent groundwater table.

21. A team of the Central Groundwater Board of the Government of India is investigating the causes of diminished water flow in the Yamuna river, and would determine how far Haryana's hundreds of heavy duty tubewells (canal-augmenting tubewells of 4 to 5 cusec capacity), dug along the river area,have been contributory factor.

These external effects emphasize the need for a systems approach to water planning, duly keeping in account the interdependence of surface and groundwater resources. Thus, in an overall model of water resources utilisation one may have to settle for exploiting groundwater resource much less intensively than that indicated purely on the basis of internal economies of investment in modern groundwater technology. To bring home this point, let D (say, 50 feet) be the economic depth of water table for tubewell technology. But at this depth, committed 'base' flow of rivers may be reduced so much so that normally 'effluent' rivers become 'influent'

PHENOMENON OF PERMANENT DECLINE IN WATER TABLE

India has a rather long tradition in maintaining records of depth of water table in certain canal tracts where introduction of canal irrigation showed early signs of water-logging. A number of observation wells were constructed with a view to keeping a close watch on any trend towards a rise in water table. Such data could be of immense value in gauging extent of decline in water table in the wake of rapid tubewell development in these canal tracts since the mid-sixties when the massive development of tubewells began. Unfortunately, the reliability of such data for the period since Independence is of doubtful value. Groundwater Directorates have recently been set up within the States' Departments of Irrigation and these have started collecting well-supervised data about depth of water table. Bulk of this data is still in the process of collection and analysis. The little that has been processed and released is not readily available. In point of fact, there is a noticeable resistance on the part of these State agencies to part with the data. This resistance is mainly attributable to political considerations, in the sense that regions experiencing decline in water table are not likely to get institutional finance for State proposals to develop more tubewells and pumpset wells in such regions. Thus, it has not been so far possible for us to find firm data about the widespread claims emanating from a number of sources, including the farmers, that "water table is going down rapidly". Some fragments of information gathered by us from some reliable or well-informed sources are presented below.

In Gujarat, which has witnessed only localized tubewell explosion in a few tracts, we have two instances of severe decline in water table. One relates to Ahmedabad city area and the other to Mehsana district. In the first case tubewells have been sunk to supplement water supply for non-irrigation uses e.g. drinking and industrial uses. For instance, the Ahmedabad Municipal Corporation raised its groundwater withdrawals from 1.70 billion gallons in 1951-52 to 18.35 billion gallons in 1971-72. According to Patel and Shah, static water levels, which were at depths of 40 to 50 feet in 1939, have now reached 165 to 170 feet in certain parts of Ahmedabad City.[22] Likewise, rapid growth of tubewells for irrigation in Mehsana district have steeply lowered the water table from a level of 40-50 feet 15 years ago to 200 feet in 1974. Equally significantly, the depth of a tubewell bore has to be 1000 feet today compared to 300 feet 15 years ago. Because of the alarming rate of about 15 feet of annual decline in water table, every user of groundwater in this area is under strong pressure to dig deeper and deeper so as to maintain his water supply.[23] According to the Irrigation Commission (1972), about 100 pumps in the district had to be either changed or lowered in 1968 alone because of steep (at some places 45 feet) recession in water table.

Punjab, Uttar Pradesh and Haryana, which have been in the forefront of tubewell explosion, are reported to have witnessed considerable fall in water table. For instance, depth of water table in Malerkotla (Punjab) is reported to have increased from 10 to 12 feet at the time of installation of State tubewells to about 30 feet now.[24] At this depth, the Persian wheel--which has been for centuries the traditional mode of groundwater lifting in Punjab plains – just cannot be used by a farmer even if his niggardly economic circumstances may make it a worthwhile proposition. Thus the disappearance of traditional wells from the Punjab agriculture may be, among other things, due to the lowering of water table to a depth much beyond the technical limits of traditional water-lifts.[25]

22. Source : Patel and Shah, (13), p. 4-85.
23. It appears that even this option of digging deeper and deeper (to prevent diminution in groundwater supply) may no longer be available to those who have already reached saline tracts undereath.Water in this area is said to be generally saline below 1000 feet.
24. The source of this information is a former Executive Engineer who had been responsible for construction and maintenance of public tubewells in Malerkotla division. According to a study by the Central Groundwater Board, the estimated groundwater extraction in 877 square km. area of Malerkotla was about 58,000 thousands hectare-metres compared to estimated replenishment of the order of 49,200 hectare-metres.
25. It is said that as a result of lowering of water table in Pubjab there has been some lowering of water table in the adjoining areas of Rajasthan State. These are those areas which had earlier benefited from rise in water table in Punjab because of water-logging caused by canals.

In Meerut district (U.P.), Daha tract is reported to have witnessed a decline in water table at the rate of two feet per year during the last 10-12 years. This has come about because groundwater extraction rate far exceeds the rate of recharge of groundwater aquifers underlying 33,000 hectare area of this tract.[26] An exercise in water balance carried out by S.P. Sangal and M.P. Bhartary has revealed that in 1972 the groundwater withdrawal in this tract was about 118,000 acre-feet compared to an estimated annual recharge of the order of 61,000 acre-feet. The approximate break-up of this groundwater withdrawal was as follows :

Private tubewells	70,210	acre-feet
Public tubewells	35,991	
Sub-total for tubewells	106,201	
Pumpset wells	333	
Persian wheels	11,113	
Total groundwater extraction	117,647	

There is no doubt that the depth of water table in this tract now exceeds the upper limit for the operation of traditional water-lifts. It was found by Sangal and Bhartary that the depth of water table, as observed from tubewell records, varied from 30 to 60 feet.[27]

In Haryana, the Directorate of Evaluation, Government of Haryana carried out a sample survey of the working of public tubewells in that State. In 1970, the average depth of water table in villages having State tubewells was significantly below the level at the time of their installation, as can be seen from the statistics below :

26. About six years ago R. N. Mathur in his Ph.D. thesis *A Study in the Groundwater Hydrology of Meerut District, U.P.; India,* (Banaras Hindu University, Varanasi, 1969) had warned that further groundwater development in this area would lower the water table.
27. Daha tract is said to have weight in U.P. political circles. Pressure is building up to improve groundwater artificial recharge in order to prevent a sharp fall in farm activity in the future. One proposal being toyed with is extension of canal network (there are no canals in the tract) of West U.P. into the area by 30 to 40 miles. Since enough water in the run-of-the-river canals of West U.P. is not available to extend canal irrigation in the area, canals will be used for recharge only during Kharif.

Depth of water table (in feet)	Number of Villages	
	at the time of installation	in 1970
0-10	2	-
10-20	45	27
20-30	14	21
30-40	-	13
40-50	2	2
50-60	-	-
60-70	1	1
	64	64

On the basis of the above data, the weighted average depth of water table works out to be 20 feet at the time of installation of the State tubewells and 24 feet in 1970. Since the installation of these 64 tubewells, the number of private tubewells and pumpset wells within the command area of these tubewells rose from just 4 to 595. In other words, the decline in water table is very likely not due to some transient factors such as deficiency in rainfall.

The Central Groundwater Board, on the basis of its exploratory tubewell studies extending over an area of about 37,000 square km. in Punjab, Haryana and West U.P., found that groundwater extraction in this part of the Indo-Gangetic basin had reached a level of about 458 thousand hectare-metres as compared to groundwater recharge of the order of 590 thousand hectare-metres.[28] Though the rate of groundwater withdrawal is only three-fourths of the rate of recharge, yet in some tracts the former had either approached or exceeded the latter, which is indicative of a trend that would force down the water table. Below are listed such tracts.

28. Based on data quoted from Dakshinamurti et al, (1b), Table 18, pp. 82-99. About 67,000 shallow tubewells, 939 public tubwells and 119,000 open wells were in the survey area.

Sl. No.	Name of the tract	Area in sq. kms.	Groundwater balance (thousand hectare-metres)		
			Recharge	Draft	Balance
1.	Kapurthala (Punjab)	1355	64.2	67.1	(-) 2.9
2.	Malerkotla (Punjab)	877	39.1	50.2	(-) 11.1
3.	Mohindergarh (Haryana)	2000	16.6	29.6	(-) 13.0
4.	Bagpat (West U.P.)	926	10.6	10.6	0.0
5.	Alipur	52	8.9	9.9	(-) 1.0
6.	Saharanpur (West U.P.)	2425	47.1	46.3	(+) 0.8

NATIONAL RESPONSE TO AND IMPACT OF PERMANENT LOWERING OF WATER TABLE

Hydrologists/hydrogeologists/geohydrologists have been in the forefront in pleading for regulating the introduction of modern tubewell technology in India so as to prevent (a) a permanent decline in water table due to excessive groundwater extraction, (b) mutual interference among tubewell owners due to lack of proper spacing among them and (c) possible irreparable damage to fresh groundwater resource found in regions close to saline sources of water. Groundwater being a State subject in the Indian Constitution, the Centre could urge the State governments to enact suitable groundwater regulation legislation. In fact, a model bill, titled "The Groundwater (Control and Regulation) Bill", was prepared five years ago by the Centre with a view to guiding the State legislatures so as to enact laws according to the particular conditions of their respective regions. Till today, no State has adopted any such statutory measure, though some States are reported to use indirect but not effective methods (such as non-grant of electric connection) to prevent further development of tubewells in some over-exploited tracts.

Before we discuss the differential impact of a permanent decline in water table on small and large farmers, as well as users of traditional and modern water-lifts, a few observations on the national response to the phenomenon of declining water table may be in order. One important economic explanation why warnings of hydrologists failed to register themselves on the legislators, as well as the users of the new groundwater technology, is the extremely low value of unit operational cost of the new technology. Consequently, lowering of water table does not give rise to serious cost implications for a farmer owing a modern water-lift. For him real cause for concern arises when the lowered water table diminishes his water yield. This eventuality, too, the adopters of new groundwater technology have, by and large, averted by measures designed to maintain their groundwater output.

Because of a number of other relieving circumstances the pinch of the much-talked about declining water table is not being felt acutely. Firstly, the recent sharp outward shift in the demand curve for irrigation water (due to the introduction of the bio-chemical revolution in the agricultural technology) has helped in cushioning the pinch of additional expenditure arising from a decline in water table. Secondly, the lowering of water table in canal-irrigated tracts has come as a welcome relief to policy makers working under the shadow of water-logging.[29] Thirdly, massive extension of rural electrification network has not only reduced a farmer's capital need for powering his tubewell/pumpset but also reduced his unit operational cost wherever he has managed to effect a switchover from a more costly oil-driven to less expensive electric power pump. Also, a farmer saves considerably on operational cost of tubewell/pumpset well in the event of a switchover from diesel to electric drive. Fourthly, continued expansion in institutional finance for farm equipment, on liberal terms, has softened the pinch of extra capital outlay that would have otherwise to be raised from village capital markets at high interest rate.

It is true that, in the matter of access to liberal finance, small and marginal farmers have not been as fortunate as large farmers backed with bureaucratic-cum-political influence. But, then, small farmers have been relatively late comers to the new groundwater technology. It is, after all, small and marginal farmers, having surplus bullock/family labour of low opportunity cost on the one hand or/and lack of access to the liberal segment of the capital finance market on the other hand, who still find using traditional groundwater-lifts a worthwhile, if not a paying proposition. Those among them, who opted for the new ground-

29. It is in Punbjab that the shadow of water-logging has been the longest. For the Bist Canal of this State it has been observed that the water table declined at the annual rate of 0.1 metre during 1965-71 whereas it was rising at the annual rate of 0.2 metre between 1941 and 1965.
Source : S.R. Sehgal and A.D. Gulati, (19), pp. 133-147.

water technology, may have been hurt badly by the decline in water table, especially if they failed to raise additional finance required for maintaining intact water yield of their wells.

It is primarily the users of high-cost traditional water-lifts who experience sizeable external diseconomies due to permanent lowering of water table by the introduction of new groundwater technology in their vicinity. Should water table recede 30 feet below ground surface, it threatens to throw them out of farming activity, unless they can effect a changeover to modern techniques of tapping groundwater. In case the water table recedes to a level which is still within the technical capability of their water-lifts, the externality shows up either as enhanced cost of lifting water or diminished water output, or both. Given that (a) technical capability of human-operated counterpoise water-lifts and (b) average cost curve of animal-operated water-lifts is far below that of the other traditional lifts, the magnitude of external diseconomies consequent upon a certain recession in water table would be far greater for owners of counter-poise appliances than for owners of animal-driven appliances.

How disastrous may be the consequences of the long-run externalities on the owners of traditional water-lifts would depend both on the magnitude of decline in water table and the ability of these owners to effect appropriate capital changes in their wells. In the event water table is lowered just below the technical limits of a human water-lift, the scope for a switchover to animal-driven devices may be exploited by deepening the dugwell. In case the water table has receded below the lifting capacity of farm animals, one may not be able to carry on farm activity without changing over to new groundwater technology. In either case, additional capital outlays are needed. It is here that the small and marginal farmers may find themselves stranded. To begin with, their lack of access to the liberal segment of the rural capital market may stand in their way. Secondly, even if they can manage to raise the additional capital required for a switchover in techniques, their size and fragmentation disabilities may be substantial enough to make the additional investment look unattractive economically. Unless they see the prospect of surmounting these disabilities in some indirect manner (say, either by selling surplus water to neighbours, or buying water from neighbours using new groundwater technology), they are doomed as peasant-farmers.

CAUSES AND CONSEQUENCES OF SHORT-RUN DEPRESSION IN WATER TABLE

Intra-aquifer external diseconomies have been accentuated by the introduction of modern tubewell technology even when (i) there is no permanent lowering of water table, and (ii) the combined extractive capacity of all the wells is less than the natural annual replenishment of the groundwater resource. Such short-run external diseconomies, as distinguished from long-run external diseconomies arising from the permanent lowering of water table, arise whenever there is crowding of wells in an aquifer. That is, wells are not properly spaced. Two wells are deemed to be properly spaced when withdrawal of water from one is unaffected by withdrawal of water from the other. Technically speaking, two wells are said to give rise to external effects, whenever their 'areas of influence' overlap.

Area of influence of a well is that part of an aquifer which supplies groundwater to the well. It is computed from 'radius of influence'(r) of a well. As water is withdrawn from a well, water held in the formations surrounding the well begins to infiltrate into the well. The surrounding formations, so affected, are in the shape of an inverted solid cone whose base is on the static surface of the water level in the well and the apex touches the surface of the water level in the well at any point of time during the course of water withdrawal or pumping. As the water level in a well recedes due to continuous operation of a water-lift, the cone becomes larger.[30] Since lowering of level in a well accelerates rate of water infiltration from surrounding formations, water level in a well gradually stabilizes. It is the radius of the base of such a stabilized cone which is defined as the 'radius of influence' of a well.

The accentuation of short-run intra-aquifer externalities as a result of the introduction of modern ground water technology *ceteris paribus* springs from the basic fact that the magnitude of 'r' for this technology is much greater than for the corresponding value for the traditional

30. An approximate determining formula for changing value of 'r' when the pumping level is receding is r = 3000 d(P) 1/2 where 'P' is the permeability coefficient of water-bearing strata and 'd' is 'drawdown' (extent of decline in water table while pumping). Source: V. I. Aravin and S.N. Numerov, *Theory of Fluid Flow in Underformable Porous Media,* 1965, p. 105.

groundwater technology. Though it is not possible to present direct evidence in support of this proposition about comparative magnitudes of radii of influence, the logic of the proposition is as follows.

As mentioned in section III of this paper, one hallmark of modern groundwater technology is its great speed in extracting groundwater from an aquifer. That is, dw/dt is of high value for modern groundwater technology, as can be seen from the following picture of comparative hourly water discharge rates for various types of water-lifts.[31]

Small power pump (5 to 6 h.p.)	7,000 gallons
Big power pump (12 to 15 h.p.)	30,000 gallons
Persian wheel	2,250 gallons
Rope-and-bucket lifts	1,800 gallons
Chain pump	1,800 gallons
Picotta	1,500 gallons
Counterpoise lifts	500 gallons

Other things being equal, 'r' is an increasing function of dw/dt. Thus, 'r' for modern groundwater technology should be deemed to be much higher than that for traditional technology. Some quantitative idea of the radius of influence and the area of influence of a tubewell in various types of water-bearing strata can be had from Table 7.4.[32]

The depression of water table within the radius of influence of a tubewell is far from uniform. The extent of this depression at a point is inversely related to the distance of that point from the tubewell pipe.

How the water table is lowered in open wells, at various distances from a tubewell, has not been empirically investigated in India. More

31. With one exception (big power pump) these discharge rates are from Jain, (7), pp. 12-13. The rate in case of a big power pump is for a deep, public tubewell of U.P. The wide range of discharge rate in the case of rope-and-bucket lifts needs remarking. The step-up in discharge rate is visualised when 3 men and 2 pairs of bullocks, instead of 2 men and one pair of bullocks are employed.
32. It may be noticed that the range of 'r' for a given soil type in this Table is rather wide, which only emphasizes the importance of many other factors e.g. whether an aquifer is confined or unconfined, time taken for the cone to stabilize, leakiness of an aquifer, etc.
Source: Murty et al, (12) p. 10.

importantly, we do not have studies indicating lowering of water table at various distances as a result of *simultaneous* operation of a few tubewells with radii of influence overlapping. In order to illustrate the potential short-run hydraulic spillover-effects on users of ground-water resources at different distances from the site of a *single tubewell*, we present in Table 7.5 some data based on pumping tests for two tubewells of West Pakistan. It is clear from such data that small and marginal farmers operating traditional water-lifts in an area having tubewells experience severe external diseconomies when tubewells depress water table in the very short-run. Such temporary depressions in water table, which are likely to occur at peak times of agricultural season when tubewells have a high probability of working simultaneously, would have deleterious impact on crops under the command of traditional water lifts. The prospect for such an eventuality is all the greater when expected rains fail. Thus, the very reliability of tubewell water during drought can be a source of undoing of farmers using traditional lifts (unless the former agrees to supply water to their doubly-stricken brethren).

If the radii of influence of tubewells overlap, the tubewell owners too would experience mutual external diseconomies in the short-run. Some of them may get less than their rated water discharges, with the result that they must prolong their pumping duration in order to raise a given volume of water. This means not only extra pumping cost but may also affect farm production at times of great water stress in crops. For some other tubewell owners temporary depression in water table may prove so deleterious that their power pumps cannot cope with the changed pumping conditions unless they either lower the pumps or instal pumps that can automatically cope with changed water table.

It has been brought to the attention of the author that heavy duty (about 500 acre-feet) canal-augmentation tubewells of the Haryana Government occasionally create serious working problems for private tubewells in their vicinity. This happens when these work in unison at peak agricultural season. Likewise, the synchronous working of private tubewells in the command of public tubewells in Daha tract of Meerut district (U.P.) adversely affects the working of public tubewells.

The foregoing account emphasizes the need for proper spacing among tubewells. This would eliminate intra-aquifer, short-run external diseconomies among tubewell owners and hopefully lessen the externalities of tubewell technology on the users of traditional modes of well irrigation. Enactment of State legislation for groundwater regulation may help in achieving far better spacing among tubewells than what prevails today or what might prevail if the tubewell explosion continues unabated. But given (a) the highly skew distribution of land holdings and (b) the lack of consolidation of land holdings, total elimination of mutual interference among tubewells may not be easy to attain even if statutory regulations are introduced.

REFERENCES

1a. Ahmed, Nazir, "Hydraulics of Tubewells', *Engineering News*, 12, December 4, 1966.

1b. Dakshinamurti, C., Michael, A.M. and Shri Mohan, Water Resources of India and Their Utilisation in Agriculture, Water Technology Centre, IARI, New Delhi, 1973.

2. Dhawan, B.D., 'Utilisation of Groundwater Resources: Public Vs. Private Tubewells', *Economic and Political Weekly,* Review of Agriculture, September 28, 1974.

3. Dhawan, B.D., "A Note on Power Requirements of Tubewells", a paper presented to Conference on Power for Growth, organised by the Punjab, Haryana and Delhi Chamber of Commerce and Industry, held at Delhi in February, 1975 and published in the "Compendium of Background Papers", published by PHDCCI.

4. Government of India, Planning Commission, Report of the Task Force on Groundwater Resources, Delhi, 1972 (mimeo).

5. Government of India, Draft Fifth Five Year Plan, Part II, Planning Commission, Delhi, 1973,

6. Government of India, Study of Tubewell Irrigation and Groundwater Development Programme, Programme Evaluation Organisation, Planning Commission, New Delhi, April 1974 (mimeo; draft report).

7. Jain, J.K., Groundwater Structures, Ministry of Agriculture, New Delhi, 1973 (mimeo).

8. Jain, J.K., Groundwater Resources, Ministry of Agriculture, 1974 (mimeo).

9. Jain, J.K., Tubewell Construction in India, New Delhi, 1972(mimeo).

10. Khan, A. Aziz, Tubewell Irrigation in Comilla Thana, Pakistan Academy for Rural Development, Comilla, Bangladesh, July 1972 (mimeo; 2nd edition).

11. Moorti, T.V., A Comparative Study of Well Irrigation in Aligarh District, India, *Cornell International Agricultural Development Bulletin,* 19, May, 1971.

12. Moorty, V.V.N., Jain K.C. and Pandey, R.N., Tubewells. Communication Centre, Punjab Agricultural University, Ludhiana, 1973.

13. Patel, A.H. and Shah, S.R., "Groundwater and Water Supply Problems of Ahmedabad", Proceedings of International Symposium on Development of Groundwater Resources, November 1973.

14. Patel, S.M. and Patel, K.V., Studies on Economics of Rural Electrification and Lift Irrigation, Indian Institute of Management, Ahmedabad, 1969.

15. Patel, S.M. and Patel, K.V., Economics of Tubewell Irrigation, Indian Institute of Management, Ahmedabad, 1971.

16. Ramanathan, K.R., Patel, A.H. and Shah S.R., "Suggestions for Augmenting the Yield of Water from the Infiltration Wells Systems at the Headworks at Dudeshwar, Ahmedabad", a paper presented at the Seventh Annual Convention of Indian Water Works Association, Bombay, held at Madras in January 1975,

17. Shukla, V.P., Interaction of Technological Change and Irrigation in Determining Farm Resource Use, Jabalpur District, India, 1967-68, *Cornell International Agricultural Development Bulletin,* 20, July, 1971,

18. Sangal, S.P. and Bhartary, M.P., Lowering of Groundwater Table in Daha Area in District Meerut, Groundwater Directorate, U.P., Technical Report, T.M. No. 5, TR (R-2) Roorkee, May 1974 (mimeo).

19. Sehgal, S.R. and Gulati, A.D, "Hydrogeological Consequences of Irrigation and Drainage Projects in the Punjab", Proceedings of International Symposium on Development of Groundwater Resources, College of Engineering, Madras 1973,

20. Zapata, Juanantonia, The Economics of Pump Irrigation: The Case of Mendoza, Argentina, Ph.D. dissertation, Department of Economics, The University of Chicago, Chicago, Dec. 1969 (unpublished).

9

Externalities of New Groundwater Technology on Small Farmers*

INTRODUCTION

Modern groundwater technology, based on power pumps, has diffused rapidly (about 17 per cent per annum) in the Indian farm sector. It is estimated that modern groundwater devices lifted about 60 per cent of the total 65 million acre-feet of groundwater lifted by all types of groundwater appliances in 1971.[1] In point of fact, the regions experiencing successful diffusion of the new bio-chemical technology also exhibit high rate of penetration of the new groundwater technology, especially tubewell technology. For example, in the States of Haryana and Punjab the traditional groundwater lifts contributed about 5 and 12 per cent, respectively, to the State's total volume of groundwater raised for irrigation in 1971, and these percentages may be even lower today.

The cost-effectiveness of the new groundwater technology vis-a-vis the traditional one has been highlighted in a number of research studies. It is not our intention to undertake a critical review of this literature. The significant aspect, to which we wish to draw attention here in the particular context of the topic of discussion in the Conference, is the exclusive concern of the researchers with demonstrating the superiority of the new technology purely from the private or individual viewpoint, without due regard to macro angle and long-term social implications. High social cost of the infrastructural capital re-

* This paper was prepared for the Indian Agricultural Economics Conference of 1975 and appeared in *Indian Journal of Agricultural Economics,* Vol. 30, No. 3, July-September, 1975

1. Government of India : Report of the Task Force on Groundwater Resources, Planning Commission, New Delhi, 1972, Annexure X (mineo.).

quired by the modern groundwater technology has not attracted appropriate attention in the benefit-cost exercises relating to the new technology.[2] More importantly, the accentuation of external diseconomies, that are invariably encountered in the unregulated exploitation of an 'ill defined-ownership' resource like groundwater, has been totally neglected while appraising the superiority of the new over the traditional technology. The likely impact of these adverse externalities on the well-being of the small and marginal farmers is the main subject of focus in this paper.

DEPTH OF WATER TABLE

In the economics of groundwater utilization the depth (D) of water table below the groundwater surface is of crucial economic significance. The operational or variable cost of lifting a unit volume of water (v) with a given technique is an increasing function of water depth, such that $dv/dD > 0$ and $d^2v/d^2D > 0$. Likewise, the unit capital cost (k) of lifting water is an increasing function of depth of water table, *i.e.*, $dk/dD > 0$ and $d^2k/d^2D > 0$. Evidently, a change in the depth of water table implies significant economic implications for an irrigator. For example, a rise in the water table would lower the marginal operational cost of groundwater, whereas a decline in the water table may not only raise the marginal operational cost but also give rise to a situation of diminished water availability, resulting in loss of farm output of the irrigator.

Fluctuations in the depth of water table occur almost continually under the complex interplay of a number of factors, of which following four are noteworthy: (1) amount of rainfall (R); (2) volume of groundwater withdrawn (W); (3) rate of recharge of ground aquifer from nearby canals, surface reservoirs and subterranean groundwater stream (I); and (4) rate of groundwater outflow into streams and drains (O).

Of central interest here is the positive relation between groundwater withdrawals and the depth of water table, i.e. for any given area dw/dt and dD/dt are *ceteris paribus* positively correlated.[3] By

2. In this connection, see this author's "Economics of Groundwater Utilisation : Public versus Private Tubewell," *Economic and Political Weekly,* Vol. IX, No. 39, September 28, 1974.
3. For a given area dD/dt is negatively associated both with dR/dt and dI/dt but has positive association with $d0/dt$.

taking both a very short and a very long view of time factor 't' one can analyse the externalities of new groundwater technology both in the short run and long run.

LONG-TERM EXTERNALITIES

The hydrologists/hydrogeologists have indeed done a signal service to the nation by alerting the policy-makers to the grave long-run consequences of excessive groundwater withdrawal with the rapid but unregulated introduction of tubewell technology. Their warnings with regard to groundwater 'mining'[4] in certain tracts experiencing rapid diffusion of the tubewell technology have not, however, been appropriately heeded in the State Legislatures.[5] One important economic explanation why these warnings have failed to register themselves on the legislators as well as the users of the new groundwater technology is the extremely low value of unit operational cost (v) of the new technology compared to the corresponding value for the traditional groundwater technology.[6] Consequently, the lowering of water table does not give rise to serious cost implications for a farmer owning a modern waterlift. For him, the real cause for concern arises when the lowered water table diminishes his water yield. This eventuality too, the adopters of new groundwater technology have, by and large, averted by measures designed to maintain intact their groundwater output.

---

4. 'Mining' of groundwater resources is deemed to occur when the quantum of water withdrawn over the entire rain cycle of good and bad years exceeds the groundwater recharge over the rain cycle period.
5. Water being a State subject, the Centre has been urging the State Governments to enact suitable groundwater legislation to prevent excessive withdrawal of groundwater by the tubewells. In this connection a model bill, titled "The Ground Water (Control and Regulation) Bill 1970" was prepared by the Centre to guide the State legislatures. So far, no State has adopted any legislative measure for groundwater regulation.
6. For the sake of illustration the dimension of costs involved, below are set out the comparative values of 'v' :

	Cost per 1000 cubic meters (Rs.)
State tubewell	1.4
Private tubewell	1.2
Persian wheel	44.6
Rope and bucket lifts (*charsa*)	93.0

Source : Based on data in T.V. Moorti : A Comparative Study of Well Irrigation in Aligarh District, India, Cornell International Agricultural Development Bulletin 19, Cornell University, Ithaca, 1970, pp. 61-64.

Because of a number of relieving circumstances the pinch of the much-talked about declining water table is not being felt acutely. Firstly, the sharp outward shift in the demand curve for irrigation water (especially for a source over which a farmer has complete control in regard to the time of application) as a result of the new bio-chemical farm technology has helped in cushioning the additional expenditure entailed by the decline in the water table. Secondly, the lowering of water table in the canal-irrigated tracts has come as a welcome relief to the policy-makers working under the shadow of waterlogging. Thirdly, massive extension of rural electrification network not only reduced a farmer's capital need for powering his tubewell/pumpset but also reduced his unit operational cost wherever he managed to effect a switch-over from a more costly oil-driven to less expensive electric power pump. Fourthly, continued expansion in institutional finance for farm equipment at liberal terms has softened the pinch of extra capital outlay that would have otherwise to be raised from the village capital market bearing exorbitant interest rate.

It is true that in the matter of access to liberal finance, the small and marginal farmers have not been as fortunate as the large farmers backed with bureaucratic-cum-political influence. But, then, the small farmers have been relatively late comers to the new groundwater technology. It is, after all, the small and marginal farmers, having surplus bullock/ family labour of low opportunity cost on the one hand or/and lack of access to the liberal segment of the capital finance market on the other hand, who still find using traditional groundwater lifts a worthwhile, if not a paying proposition. Those among them, who opted for the new groundwater technology, may have been hurt badly by the decline in the water table, especially if they failed to raise additional finance required for maintaining intact water yield of their wells.

It is primarily the users of high cost (high variable cost though low capital cost) traditional waterlifts who experience sizable external dis-economies due to permanent lowering of water table by the introduction of new groundwater technology in their vicinity. In case the depth of water table recedes to a level which is still within the *technical capability* of their waterlifts, the externality shows up either as enhanced cost of lifting water or diminished water output, or both. Given that (a) technical capability of human-operated counterpoise waterlifts (*Dhenkali, Lot, Tera, Shadoof,* etc.) is almost half that of animal-operated waterlifts (*Rahat, Mhote, Charsa, Kapila* and *Pun*) and (b) animal-operated waterlifts are usually less costly than the other traditional lifts, the magnitude of external diseconomies consequent upon a certain recession in the water table would be far greater for the owners of counterpoise appliances than for the owners of animal-driven appliances. In point of fact, the counterpoise lifts practically become inoperative when the depth of water table crosses the 15-feet limit. The corresponding limiting value of 'D' for animal-driven devices is about 30 feet. Beyond these critical limits of 'D', permanent lowering of water table threatens to throw out of farming activity the users of traditional modes of groundwater lifting unless they can effect a change-over in their techniques of tapping groundwater.

How disastrous may be the consequences of the long run externalities on the owners of traditional waterlifts would depend upon both on the magnitude of decline in the water table and the ability of these owners to effect appropriate capital changes in their wells. In the event the water table is lowered just below the technical limit of a human waterlift, the scope for a switch-over to animal-driven devices may be exploited by deepening the tubewell. In case the water table has receded below the lifting capacity of farm animals, one may not be able to carry on farm activity without changing over to new groundwater technology. In either case, additional capital outlays are needed. It is here that the small and marginal farmers may find themselves at a great disadvantage. To begin with, their lack of access to the liberal segment of the rural capital market may stand in their way. Secondly, even if they can manage to raise additional capital required for a switch-over in techniques, their size and fragmentation disabilities may be substantial enough to make the additional investment look unattractive uneconomically. Unless they see a prospect of surmounting these disabilities in some indirect manner (say, either by selling surplus water to the neighbours, or buying water from the neighbours using new groundwater technology), they are doomed as peasant farmers.

SHORT-TERM EXTERNALITIES

Adverse short run externalities of the tubewell technology on the users of traditional waterlifts and open wells fitted with power pumps -- to be distinguished from short run externalities which the tubewell owners impose on one another due to lack of proper spacing among the tubewells -- are not so well realised in the literature as are the long-term externalities of the unregulated introduction of the tubewell technology. In order to bring out these externalities it is necessary to dwell here on induced changes in the water table brought by pumping in the vicinity of a tubewell. As water is pumped out, water held in the formations surrounding the tubewell begins to infiltrate into the tubewell. The surrounding formations so affected are in the shape of a solid cone whose base is on the surface of the static water level and the apex touches the surface of the pumping level of groundwater. The radius of the circular base of this inverted cone is known as the 'radius of influence' ('r') of the well, and as the pumping level recedes in the well 'r' becomes larger.[7]

The value of 'r' is vitally dependent on two factors : groundwater withdrawal rate (dw/dt) and the physical characteristics of the water-bearing strata. dw/dt is of high magnitude for modern groundwater technology compared to the traditional technology, with the result that the area of influence of the former technology is several times that of the latter technology. Thus, compared to 5 to 8 acre-feet of water withdrawn annually through a traditional open well, 12-20 acre-feet is raised by an open well fitted with a power pumpset; 15-40 acre-feet by a shallow tubewell and 200-300 acre-feet by a deep State-owned tubewell.[8] Likewise, the comparatively high speed of the modern groundwater technology may be gauged from the fact that a power pump can almost empty an open well in just 2 to 3 hours of operation whereas a traditional waterlift might take 10 to 12 hours of operation for emptying the same well.

Within the radius of influence of a well the extent of adverse externality -- as reflected in lowering of water table -- experienced by another groundwater exploiter is in inverse proportion to his distance from the well in question. How the water table is lowered in open wells at various distances from a tubewell has not been empirically and thoroughly investigated in India.[9] More importantly, we do not have

7. The determining formula for changing value of 'r' when the pumping level is receding is r = 3000 d (P) 1/2 where 'P' is the permeability coefficient and 'd' is 'drawdown' (extent of decline in water table while pumping).
Sources : V.I. Aravin and S.N. Numerov : Theory of Fluid Flow in Underformable Porus Media, 1965, p. 105.
8. *Source* : Reference to footnote 1.
9. The author is aware of only few empirical studies conducted by the U.P. Irrigation Research Institute, Roorkee. However, these are based on just one observation well in the vicinity of a State tubewell.

studies indicating the lowering of water table at various distances as a result of *simultaneous* operation of a few tubewells with radii of influence overlapping. In order to illustrate the potential short run hydraulic spillover-effects on users of groundwater resource at different distances from the site of a *single tubewell,* we present in Table 7.5 some date based on pumping tests for two tubewells of West Pakistan. It is clear from such data that the small and marginal farmers operating the traditional waterlifts in an area having tubewells may be experiencing external diseconomies when tubewells depress the water table in the very short run. Such temporary depressions in the water table, which are likely to occur at the peak time of agricultural season when tubewells have a high probability of working simultaneously, may have deleterious impact on crops under the command of traditional waterlifts. The prospects for such an eventuality is all the more when the expected rains fail. Thus, the very reliability of tubewell water in drought time can be a source of undoing of farmers using traditional lifts (unless the former agree to supply water to their doubly-stricken brethren).

10

Underutilisation of Groundwater Resources

A Case Study of East Uttar Pradesh*

I

Groundwater endowment of east Uttar Pradesh is immense. Equally noteworthy are the favourable conditions under which groundwater occurs. Depth to water table, a crucial determinant of cost for exploiting groundwater, is so low at places that groundwater can be raised to the surface with wholly human-operated water-lifts. Groundwater is found in a series of (sandy) aquifers, separated by clay lenzes, which can yield a sustained supply of sweet water, especially when a dug-well is strengthened with masonry work and through the bottom of its pit are lowered pipes to interconnect two or more water-bearing strata. However, if instead of such semi-modernisation of a traditional well the fully modern technique of tubewell irrigation is employed in tapping these aquifers, copious supplies that are largely unaffected by down swings in the rain cycle can be produced for irrigation. It is this copiousness or perenniality that makes not only irrigated cropping possible in the region during all the three crop seasons but also confers on groundwater resource a key role in the practice of HYV farming with its exacting demand for irrigation.

In terms of volume of water, the groundwater potential of east UP can be placed at about 2.5 million hectare-metres, a quantity that can be safely withdrawn year after year. For, underground reservoirs are replenished annually to this extent through infiltration of rain (20 to 25 per cent of rainfall), seepage from unlined canals of the region, and percolation of irrigation water applied to the fields. At present,

* This paper is part of a larger study, entitled Groundwater Irrigation in East U.P. (IEG, mimeo, 1978). It was published in *Economic and Political Weekly* (Review of Agriculture), September 1980.

barely two-fifths of the regional potential has been exploited. Were all the groundwater fully tapped during 1975-76, it might have sufficed for about 4.5 out of 6 million hectares of cultivable area of east UP. In short, the total water resources of east UP are quite adequate for supporting triple or year-round cropping.

As elsewhere in India, the principal demand for irrigation here is during the rain-deficient Rabi season – 85 per cent of the irrigated area of the region is Rabi area. Monsoon rains (normal annual of about 110 cms) are adequate for kharif cultivation, with the possible exception of kharif paddy for which intensive cultivation without irrigation facilities at the sowing and maturing stages of the crop is risky. There is not much 'zaid' (summer) cultivation in the region – 'zaid' crops accounted for hardly 3 per cent of the total gross cropped area -- because summer cropping is possible only with a copious source of water. On the whole, about 35 per cent of the total cropped area of east UP is protected by irrigation, of which the share of groundwater sources is six-tenths.

According to official statistics, there were a little over 625,000 masonry wells and about 181,000 private tubewells in east UP at the end of March 1977. Since these statistics are built on a progressive basis – additions within a time period are simply added to the cumulative number at the beginning of this period –, these are not adjusted for structures going out of use, either due to lack of major repairs or simple abandonment in the wake of some alternative source of irrigation becoming available to their owners. Hence such statistics exaggerate the number of groundwater structures in use even if one may accept the underlying data with regard to the (gross) annual additions to such structures, compiled on the basis of reports filed by the district officers who in turn rely on the records of the Block Development Offices. A census count in 1972 brought to light an excess recording of the order of 45 per cent in the case of dug wells.

PUBLIC TUBEWELL IRRIGATION

Uttar Pradesh is a pioneer state where tubewell irrigation was introduced more than 40 years ago. Till 1950, state tubewells were established in west UP only. With the advent of planning, east UP started receiving due attention in the matter of public tubewells, for which its case was strengthened by the fact that this region did not have its due share of public canals established during the pre-Independence days. The number of state tubewells, each with its own independent command area and an approximate rated capacity of 1.5 cusec, stood at 5,879, which constituted 46 per cent of the total number of state tubewells in the entire Uttar Pradesh. Their estimated irrigational potential was of the order of 0.637 million hectares, though the actual irrigated was probably less. (The state and private tubewells together irrigated about 0.9 million hectares during 1975-76).

There is as yet no firm analysis about the economic performance of public tubewells of east UP. The general impression is that the private benefit-cost ratio of the beneficiaries is above unity. For instance, the Planning Research and Action Institute (PRAI), Lucknow, placed this B-C ratio on the basis of a sample survey of beneficiary farmers in the command of ten state tubewells in Varanasi district at 2.20. This relates, however, to 'good tubewells', each having a working duration of more than 4,500 hours during 1966-67, a drought year. More recent estimates of social benefit-cost ratio are not available.

The financial working of the state tubewells is dismal. Annual expenditure, including depreciation charge but excluding interest charge, exceeded revenue receipts of all state tubewells by 120 per cent, and a little over 200 per cent if interest cost is also included in the resource cost. The annual financial loss per state tubewell came to Rs. 7,500 (Rs. 12,850 if interest cost is included) during 1976-77. (This is about nine per cent of the capital invested in state tubewells). There is no ground for believing that the financial performance of state tubewells located in east UP would be in any way different from the overall picture for the entire state.

DUG WELL IRRIGATION

Dug well irrigation constituted the principal source of irrigation in east UP till the tubewell spurt of the 1970s. If one corrects for the discrepancy between the official estimate and the census count of 1972, the number of dug wells would be about 436,000 in 1977. If one deducts also the dug wells which were found to be not in working condition during the 1972 census, the number of working wells may be placed at 412,000, a figure which is marginally above the number reportedly existing in 1951. One may thus conclude that the number of dug wells has not changed materially during the period of planning, though most of the wells are now 'pucca' (masonry) wells, some with boring at the bottom. 'Kutcha' wells, which at one time were of considerable importance in this region, are no longer favoured by the farmers because of their short life (one to two years), enhanced risk of caving in if the depth is increased with a view to making water supply less vulnerable to drought when water table recedes; their inability to meet the exacting water needs of HYV crops especially during Rabi and 'zaid' seasons, etc. These disadvantages probably outweighed their advantage of being extremely low cost constructions in the proverbially capital-scarce economy of east UP.

The salient characteristics of waterlifts used by east UP farmers is revealed by the PRAI census report of 1972. Firstly, only 4 per cent of the wells were equipped with power pumpsets, of which three-fourths were diesel sets. Secondly, animal-operated water-lifts in contrast to the situation in west UP where 'rahat' (Persian wheel) was the most common animal-operated water-lift. The comparatively greater reliance on human-operated water-lifts and 'charsa' in east UP is in harmony with the relative resource endowment of the two regions. For, both human-operated water-lifts and 'charsa' are more labour-using and less capital-using as compared to the Persian wheel. For example, the capital cost in case of a 'charsa' came to Rs. 150 as compared to Rs. 2,000 for a Persian wheel in 1975. The capital cost of human-operated lifts, like 'dhenkali', would be still lower than that of a 'charsa'. More importantly, the purchased or cash component of the capital expenditure in the case of a 'dhenkali' or 'charsa' is relatively smaller than that of a Persian wheel.

87 per cent of the dug wells in east UP were self-financed by the owners, and the government taccavi loans financing the construction of another 11 per cent of the wells. The marginal role of institutional agencies is understandable because 85 per cent of the dug wells were constructed prior to the mid-sixties, when little institutional finance was available for acquiring farm assets. Precisely at this time the tubewell technology became a relatively more attractive proposition for the medium and large farmers who then primarily fulfilled the

criteria for institutional term loans. Lately, the trend in gross annual additions in wells has been unmistakably downwards. Admittedly, some of these additions since 1951 (about quarter million of masonry wells) could be for the replacement of old wells which had outlived their lives. But it is known that well irrigation is abandoned on a considerable scale when cheap public irrigation becomes available. In fact, the introduction of Sarda Canal in the central UP led to such a pervasive abandonment of well irrigation that the principal benefit from the canal turned out to be reduction in irrigation cost for the farmers, and not net expansion of area under irrigation.

There is a noteworthy trend in respect of modernisation of the water-lift as well as the dug well through well-boring. In the beginning, Persian wheel, which was the mainstay of well-irrigation in the Indus and the west Gangetic plains till the advent of tubewells, attracted the attention of east UP farmers, but after the mid-sixties power-driven pumpsets, whose operating costs are low, proved more attractive. In all, the number of Persian wheels rose from a little less than 800 in 1951 to over 61,000 in 1977. Likewise, the number of power pumpsets soared from 320 to about 97,000. (A cross-check with the PRAI census reveals substantial overstatement – 37,000 pumpsets as against the progressive number of about 60,000.) Table 10.1 presents Plan-wise annual additions of Persian wheels and pumpsets in east UP.

TABLE 10.1

Annual Addition of Pesian Wheels and Pumpsets in East UP

	Persian Wheel	*Pumpset*
I Plan	91	5
II Plan	427	189
III Plan	4,825	1,502
Three Annual Plans	5,034	10,664
1969-70	1,391	6,853
1970-71	1,870	6,538
1971-72	1,194	5,644
1972-73	682	5,725
1973-74	484	8,457
1974-75	732	8,397
1975-76	491	7,905
1976-77	258	6,760

The more recent trend is an unmistakable tapering off in the expansion of Persian wheels. While the recent annual additions to the stock of pumpsets is steady and substantial, one is not certain if it is mainly due to pumpset demand for dug well irrigation. More importantly, a substantial part of this annual demand could be for replacement purposes because the life of a pumpset is rather short (5 to 8 years).

PRIVATE TUBEWELL IRRIGATION

If the statistics of the state's minor irrigation department are any guide, east UP is not far behind the rest of the state in the matter of private tubewell irrigation. As on March 31, 1976, the number of private tubewells stood at 28 per 1,000 hectares of cultivated area of the region. This compares favourably with the corresponding state average of 32.

The three major sources of finance for private tubewell investment in east UP are government taccavi loans, land development bank and self-finance. Whereas self-finance was the single most important source of funding in west UP, government loans had that position in the case of east UP, and land development bank in the case of central UP. More specifically, for 57 per cent of the tubewells in west UP, 34 per cent in east UP and 36 per cent in central UP, the principal source of finance was the farmers' own resources. This confirms the general impression that capacity to self-finance farm investments diminishes as one moves from the western Indo-Gangetic plains to the eastern Gangetic plains. Till the beginning of the 1970s, government more than the institutional agencies such as land development banks, commercial banks and co-operative societies had played the role of bridging the saving gap in east UP. (This composition of financing would be altered in favour of institutional sources, particularly UP Land Development Bank, if due account is taken of the private tubewells established since 1972. Ten districts of east UP have been the principal beneficiaries of the World Bank Loan of over Rs. 55 crore, of which Rs. 51 crore had been disbursed by June 1977, bulk of which was absorbed in the construction of tubewells in which farmers' own share was about 10 per cent of the total investment). If one were to assume a life of 10 years for a private tubewell and deduct the replacement demand from the gross additions to the stock each year, the picture of net annual addition to the tubewell stock reveals a 'parabolic' pattern of growth during the 1970s (Table 10.2).

TABLE 10.2

Net Annual Addition to Tubewell Stock in East UP

1966-67	4,118	(4,175)
1970-71	11,867	(11,966)
1971-72	15,213	(15,571)
1972-73	19,742	(20,265)
1973-74	25,562	(26,329)
1974-75	22,781	(23,905)
1975-76	21,203	(22,848)
1976-77	16,436	(20,611)

Note : Figures in brackets are official estimates of gross annual addition.

The peak in the tubewell spurt is most probably now over in east UP, and almost one-fifth of the reported gross additions to the stock now is for replacement demand. The decline in private tubewell construction during the last three years is probably not so much due to any change in the profitability of a tubewell investment; it is that the structure of landholdings in east UP cannot sustain the expansion in the number of private tubewells. If the structure of landholdings, as revealed by the Agricultural Census of UP for 1970-71 is any guide, one can say that less than 10 per cent of the operated holdings in east UP were above the minimum economic land size (about 5 acres) for an investment in a tubewell in 1975. As this pool of holdings is gradually exhausted, the annual net additions to tubewell stock are bound to diminish. This deceleration can be conceivably arrested if (1) objective conditions for investment in a tubewell improve so as to lower the minimum economic farm size; or (2) if farmers develop a spirit of co-operation so as to surmount their individual size or land fragmentation disability for investment in a tubewell.

OVERALL GROUNDWATER IRRIGATION

In spite of substantial expansion in tubewells and water yield-raising modernisation of dug wells, the growth in area irrigated by groundwater during the planning period has been modest, from 1.2 million hectares in 1950-51 to 1.5 million hectares in 1975-76, that is about 25 per cent in as many years. No doubt, the bulk of this growth occurred during the ten-year period 1965-66 to 1975-76 (Table 10.3). There was no improvement in the share of groundwater irrigation in overall

TABLE 10.3

Progress of Irrigation in East UP

('000 hectares)

Year	*Gross Irrigated Area*	*Net Area Irrigated by*		*A as a Percentage of B*
		Groundwater Sources (A)	*All Sources(B)*	
1950-51	1,922	1,202	1,880	64
1955-56	2,089	1,290	2,030	64
1960-61	2,094	1,281	2,011	64
1965-66	2,196	1,295	2,094	62
1970-71	2,493	1,418	2,246	63
1975-76	2,704	1,520	2,476	61

irrigation during 1950-75. The slight downward drift in this share, as revealed in column 4 of Table 10.1, need not be taken at its face value. If one were to take due note of the fact that each 'net' acre of groundwater-irrigated area had 7 percentage points higher intensity of irrigation in 1975-76 than in 1950-51, it is quite probable that groundwater improved its effective share in overall irrigation during the above period. The overall expansion in irrigation facilities in east UP since the beginning of planning in 1951 has been slow, with an annual growth rate of the order of one and quarter per cent. The net area irrigated by wells/tubewells rose at an annual (semi-logarithmic) rate of about 0.8 per cent during 1950-51 to 1972-73, a rate which was statistically no different from the corresponding rate of 0.86 per cent for net area irrigated by all sources. While these rates of expansion, taken alongwith the qualitative change in the composition of groundwater sources, do indicate a step-up in the sluggish growth of the pre-planning era – 0.3 per cent per annum in groundwater-irrigated area and 0.2 per cent in total net irrigated area during the first half of the twentieth century – these are far from the desired rates for a predominantly unirrigated agriculture that is striving for the adoption of HYV strategy. Why the rates are low even during a period of planned development is an extremely important question in the particular context of underutilisation of groundwater resources. In fact, despite a substantial increment to the reported stock of tubewells (public as well as private ones) and the improvements in masonry wells in east UP over the last 27 years, there has been no commensurate expansion in the irrigated area.

Were we to adopt the norms of irrigation potential of the state's minor irrigation department – 8 hectares, 3.2 hectares and 1.2 hectares per private tubewell, pumpset and masonry well respectively – the additional irrigation potential created through private sources of groundwater irrigation in east UP during the planning period 1951-77 would come to an impressive total of over 1.63 million hectares. Adding to it an additional potential of over 0.6 million hectares created through about 5,000 state tubewells during the same period the overall addition to groundwater potential in east UP would rise to about 2.25 million hectares, a figure which is almost a million hectares above the groundwater-irrigated area of the region at the beginning of 1950-51. If this additional potential were fully utilised (as per the norms) and the whole of it is deemed as pure increment to the baseline (1950-51) figure of 1.2 million hectares, the compound rate of growth of groundwater-irrigated area of entire east UP would have been about 4 per cent per annum, a rate that is almost four-fold the rate actually observed. Such a wide difference between the actual and the potential rates is evidently indicative of either large-scale replacement of the old capacity by the newly-created capacity, or sharp decline in the level of utilisation of capacity, or both. While it is known that dug well irrigation tends to decline in canal tracts, an additional source affecting this mode of irrigation adversely is the substantial rise in tubewell irrigation, both on public and private account. If the statistics of expansion in canal irrigation in east UP (Table 10.4) are any guide, one might attribute a possible reduction of about 0.5 million hectares (as compared to an additional potential of 2.25 million hectares) in dug well irrigation to new canal irrigation introduced in east UP since 1950-51.

TABLE 10.4

Net Area Irrigated by Canals in East UP

(hectares)

1950-51	93,000
1955-56	118,000
1960-61	233,000
1965-66	319,000
1970-71	425,000
1971-72	NA
1972-73	384,000
1973-74	NA
1974-75	495,000
1975-76	536,000

The foregoing analysis brings out highly non-additive character of irrigation potential to the area under irrigation existing at the time of commencement of planning. This raises a number of questions. Is each irrigated acre now receiving far more water than in traditional farming of the early fifties? Does introduction of public irrigation still cause considerable abandonment of well irrigation in high rainfall areas? Is depreciation of irrigation works taking place faster? Or, do the statistics have serious biases – upward bias in the number of groundwater devices installed as well as the official norms about their irrigation potential and downward bias in the statistics of area irrigated?

II

For purposes of evolving an appropriate public policy for groundwater development in the east Gangetic plains in general and east UP in particular, one needs quantitatively firm answers to the following questions. What are the current private costs and returns from various modes of groundwater irrigation? How do they compare with the corresponding costs and returns of the surface irrigation works? And how do these change with farm size? How does the private rate of return on a given investment in a given mode of groundwater irrigation compare with the reportedly high interest rates ruling in the rural capital market? How far is this return altered with the availability of institutional loans at low interest rate, both for long-term and short-term financing of investments? How do these private returns compare with social returns assessed within the standard social benefit-cost framework? Finally, how does the social B-C ratio for public tubewells compare with the social B-C ratio for private tubewells as well as public canals ?

INADEQUATE FIELD SURVEYS

Unfortunately, quantitative answers to these questions are not readily available. To begin with, few field surveys have been conducted to ascertain the returns from well irrigation. Since the accounting method is the only appropriate method that can yield reliable information about the costs and returns of well irrigation, field studies in groundwater irrigation become time-consuming and expensive, with the result that sample size tends to become small for individual scholars with small funds. Consequently, one generally gets an average picture of costs and returns for a few broad categories of wells such as electric tubewells, diesel tubewells, dug wells with traditional water-lifts and dug wells with pumpsets. Not only analysis according to the size characteristics of a given category of well is not feasible with a small sample, but analysis according to the farm characteristics (such as farm size, its extent of consolidation in space, and sharing of well water with other farms) is often not attempted.

More importantly, inter-farmer differences in valuation of inputs in well irrigation and benefits from well irrigation have not been treated appropriately in the benefit-cost analysis conducted on survey data. On the cost side, the capital cost of a well is taken at historical prices, that is, cost at installation time. Since then, construction costs may have risen sharply because of inflation. The larger the proportion of such old wells vis-a-vis new wells in a survey data, the greater the upward bias in the B-C ratio for the sample wells. The danger of such upward bias is now high because in any random sample many wells constructed at half the cost of today would be chosen because of substantial expansion in wells during the inflationary period of the last 10 years.

Imputation of values on a uniform basis to two inputs of well irrigation, namely, labour and interest on capital sunk in a well, conceals substantial inter-farmer differences in cost of well irrigation. Quite often, these are arrived at by the field investigators by applying ruling wage or interest rate to the reported labour input in physical terms and the capital cost at historical prices for every farmer who often cannot report these costs in actual terms. This uniform procedure of valuation distorts both the inter-farmer as well as inter-well differences in cost of irrigation.

The large inter-farmer differences in benefits from a unit of well water are today primarily due to the extent of adoption of HYV farming. Evidently, adoption of HYV farming is determined by the type of well irrigation, a farmer's own drive and initiative, the quality of and access to agricultural extension services and inputs needed for HYV agriculture. Thus overall low benefits from well irrigation that arise from inadequate extension services and short-term credit for the inputs needed for HYV agriculture in east Gangetic plains can create a situation of low returns from irrigation.

In the absence of any field data even of a small sample size, we propose to construct in what follows some hypothetical estimates of costs and benefit from different modes of groundwater irrigation to various categories of farmers in east UP. The estimates of both costs and benefits are admittedly notional. But these are underpinned by a set of working assumptions which are not unrealistic or untenable as may appear. To be more precise, the cost estimates presented here are nearer the current reality in east UP whereas estimates of benefits are nearer the potential benefits that can be realised from the current farm technology.

ESTIMATES OF COSTS

Well-wise estimates of capital cost, operational cost, and repair and maintenance cost are derived from a very small sample survey (unpublished) conducted by the Evaluation Division of the UP State Planning Institute in the districts of Varanasi, Banda and Meerut. The three cost components of this 'base' data (human labour, bullock labour and interest) have been revalued according to the category of farmers who have been divided into seven categories – three of large farmers and four of small farmers – for each of which a price set has been chosen (Table 10.5).

TABLE 10.5

Selected Price Sets for Large and Small Farmers

	Farmer Category	*Daily Wage Rate (Rs.)*	*Daily Cost of a Pair of Bullock (Rs.)*	*Annual Interest Rate (Per Cent)*
I:	*Large farmers*			
	Category (1)	3.75	5	14
	Category (2)	3.75	5	14
	Category (3)	3.75	5	28
II.	*Small farmers*			
	Category (4)	0.50	5	14
	Category (5)	0.50	5	56
	Category (6)	2.50	5	14
	Category (7)	2.50	5	56

Among the large farmers, the category (1) is probably the most numerous in east UP, whereas the category (2) is envisaged to represent farmers with tractors, which raises for them the cost of maintaining bullocks for lifting water from dug wells. Category (3) represents large farmers without tractors and largely dependent on their own resources for meeting capital costs of well irrigation but the opportunity cost of their capital funds is higher than the lending rate of the institutional agencies but less than the rate charged by the money-lenders. In the case of small farmers, only unit wage cost and interest cost are being varied while the unit bullock cost is pegged at the extra feed and depreciation cost for using the plough bullocks for water-lifting. The categories (4) and (5) represent small farmers with a perception of very low value for their own labour used in well irrigation but with sharp differential in their access to the institutional funds for investment in well irrigation. On the other hand, the categories (6) and (7) of the small farmers -- which may be less numerous than the other two categories of small farmers in east UP – represent farmers with considerable opportunity cost of their time but somewhat below the wage rate for hiring labour for well irrigation.

Investment in well irrigation being an indivisible investment, the unit cost of irrigation, say an acre-inch of water, is an inverse function of the level of utilisation of a well/tubewell. Thus inter-mode comparisons can be meaningful at given levels of water output from the various modes of irrigation. In the case of traditional wells, the cost-benefit analysis relates to two levels of output, namely 1.2 and 2.2 million gallons of water per annum. An output of 2.2 million gallons corresponds to about 100 acre-inches of water, which is probably the capacity output of a dug-cum-bore well with traditional water-lift, and which should suffice for a three-acre farm irrigated and cropped with a crop intensity of about 200 per cent. In the case of a dug well fitted with a pumpset, the benefit-cost analysis is done for water output of 2.2 and 2.9 million gallons. Here, 2.9 million gallons is the annual capacity output of a dug-cum-bore well operated with a pumpset that should suffice for a four-acre farm with a crop intensity of 225 per cent. It is in the case of tubewells that six levels of output are being considered : 1.2, 2.2, 2.9. 6.5, 11.5 and 23.0 million gallons, of which the last one is very near the rated potential capacity of a tubewell. These levels correspond to 100, 180, 240, 540, 960 and 1,920 running hours per tubewell per annum.

For the year 1975, the average value of benefits from private tubewell irrigation for east UP worked out to Rs. 539 per crop acre, which was an average of the following four crops : HYV wheat Rs. 821 (.40); Tall or 'desi' wheat Rs. 290 (.10); Sugarcane Rs. 669 (.15); and Kharif rice Rs. 233 (.35). (Figures in brackets are weights assumed for crops in the farm having a private tubewell). If one assumes 30 inches of irrigation for sugarcane, 12 inches for HYV wheat and Kharif rice and 6 inches for 'desi' wheat, the benefits per acre-inch of tubewell water for 1975 in east UP would come to about Rs. 39. Our task now is to vary this amount of benefits according to the category of farmers as well as the mode of well irrigation.

There are six modes of well irrigation : (1) masonry dug well equipped with rope-and-bucket lift ('charsa'); (2) masonry dug well equipped with Persian wheel ('rahat'); (3) masonry dug well fitted with electric pumpset of 5 HP; (4) masonry dug well fitted with diesel pumpset of 5 HP; (5) 5 HP electric tubewell; and (6) 5 HP diesel tubewell.

The task of assessing benefits from an acre-inch of groundwater for various categories of farmers operating different modes of groundwater irrigation is being settled here in an *ad hoc* manner. For this purpose, it is reasonable to presume that (1) large farmers derive more benefit from irrigation than do small farmers; and (2) farmers owning tubewells derive more benefit than do farmers owning dug wells. The underlying forces behind the differential (1) are well known to scholars of Indian agriculture. These are : some discrimination against small farmers both in the product and factor markets; lower risk-taking capacity of the smaller farmers; lower access to agricultural extension services; etc. A differential of Rs 5 per acre-inch of water is being assumed on this count, and a further differential of Rs 5 is being taken for the situation where small farmers are forced to cut back the consumption of fertilisers because of high interest rate. For the sake of simplification, a differential of Rs 5 per acre-inch is being visualised in the case of tubewell owners vis-a-vis dug well owners. The basis for this differential is the perenniality and greater realiability of tubewell water vis-a-vis dug-cum-bore wells as a result of which tubewell owners can take up more remunerative crop pattern.

In view of these considerations, the following values of benefits per acre-inch of groundwater have been used in assessing comparative benefit-cost- ratios for large and small categories of farmers in east UP.

Large farmers :	
with tubewells	Rs 45
with dug wells	Rs 40
Small farmers :	
with tubewells	Rs 40 (Rs 35 in case of high interest rate).
with dug wells	Rs 35 (Rs 30 in case of high interest rate).

COMPARATIVE UNIT COST OF IRRIGATION

Unit cost of irrigation pertains to an acre-inch (22,500 gallons) of water. Table 10.6 shows that it varies considerably with well type, farmer category and level of utilisation of a well. If one assumes that inter-farmer and inter-well differences in benefits from an acre-inch of water are of the same magnitude for a given level of water output per annum, the unit cost statistics would be a good guide to farmers' preferences among the various modes of groundwater irrigation, Proceeding on this presumption the following points emerge.

TABLE 10.6

Estimates of Unit Cost of Irrigation

Types of Well	*Annual Water Output (m. gallons)*	*Large Farmers of Category* 1	2	3	*Small Farmers of Category* 4	5	6	7
Well with 'charsa'	1.2	37	65	45	16	26	28	41
	2.2	30	59	35	13	18	20	34
Well with 'rahat'	1.2	33	53	46	19	41	28	61
	2.2	24	43	31	13	25	20	38
Well with electric	2.2	68	68	93	55	111	64	130
pumpset	2.9	28	28	38	22	45	27	55
Well with diesel	2.2	44	44	63	33	77	40	91
pumpset	2.9	37	37	51	27	61	33	64
Electric tubewell	1.2	63	63	91	57	131	61	141
	2.2	37	37	53	32	74	36	80
	2.9	29	29	40	25	56	27	61
	6.5	16	16	21	13	26	15	29
	11.5	9	9	12	7	15	8	16
	23.0	5	5	6	4	8	4	8
Diesel tubewell	1.2	72	72	108	61	160	68	173
	2.2	46	46	66	36	92	41	99
	2.9	38	38	52	21	70	34	77
	6.5	26	26	32	14	36	23	42
	11.5	20	20	24	10	23	17	28
	23.0	16	16	18	8	14	13	19

In east UP, no single mode of well irrigation dominates over other modes for all class of farmers and for all levels of water output. This explains why fresh investments in private groundwater irrigation are taking place in all kinds of wells. For example, as many as 4,491 masonry wells, 20,611 private tubewells, 6,760 power pumpsets and 258 Persian wheels were reportedly installed in the region during 1975-76. Small farmers find traditional modes of well irrigation less costly – more so if they do not hope to utilise the full rated capacity of wells – than the semi-modern or wholly modern modes of groundwater irrigation. For them, tubewell would become a better option if (i) they get access to loan funds at low rates of interest and (ii) they can effect substantial sale of surplus tubewell water to their neighbouring farmers.

Not unexpectedly, the large farmers do find tubewell irrigation less expensive than traditional or semi-modern modes of groundwater irrigation, but only so long as their demand for water is much above the rated capacity output of the latter. If for some reason -- say fragmentation of holdings – their demand is just equal to the rated capacity of a dug-cum-bore well fitted with a traditional water-lift, large farmers of category (1) may not ignore traditional modes vis-a-vis tubewell irrigation whose unit cost is very sensitive to the level of utilisation of a tubewell.

For the sake of comparison, it would be useful to record here the unit cost of irrigation from public tubewells. For a 1.5 cusec tubewell, this cost is extremely low when the tubewell is utilised intensively, say for 4,000 hours in a year. In the case of unavoidable power tariff that is currently in vogue in UP, the cost per acre-inch could be about Rs. 5 which is indeed far below the corresponding cost for traditional and semi-traditional wells for all categories of farmers. Only private electric tubewells compare favourably, provided they too work very near their potential capacities. In the event of underutilisation of public tubewells, their unit cost of irrigation would rise.

POTENTIAL BENEFIT-COST RATIOS

Assuming that the benefits from an acre-inch of water are invariant with respect to the level of utilisation from a well, the hypothetical estimates of benefit-cost ratios for various categories of farmers from different types of wells are presented in Table 10.7. One notices wide inter-farmer and inter-well differences in B-C ratios, which when less than unity are indicative of the economic infeasibility of that option for a particular category of farmers.

TABLE 10. 7

Estimates of Benefit-Cost Ratios for Wells in East UP

Type of Well	*Annual Water Output (million gallons)*	*Large Farmers of Category*			*Small Farmers of Category*			
		1	*2*	*3*	*4*	*5*	*6*	*7*
Well with 'charsa'	1.2	1.1	0.6	0.9	2.2	1.2	1.3	0.7
	2.2	1.3	0.7	1.1	2.7	1.7	1.8	0.9
Well with 'rahat'	1.2	1.2	0.8	0.9	1.8	0.7	1.3	0.5
	2.2	1.7	0.9	1.3	2.7	1.2	1.8	0.8
Well with electric	2.2	0.6	0.6	0.4	0.6	0.3	0.5	0.2
pumpset	2.9	1.4	1.4	1.1	1.6	0.7	1.3	0.5
Well with diesel	2.2	0.9	0.9	0.6	1.1	0.4	0.9	0.3
pumpset	2.9	1.1	1.1	0.8	1.3	0.5	1.1	0.5
Electric tubewell	1.2	0.7	0.7	0.5	0.7	0.3	0.7	0.2
	2.2	1.2	1.2	0.7	1.3	0.5	1.1	0.4
	2.9	1.6	1.6	1.1	1.6	0.6	1.5	0.6
	6.5	2.8	2.8	2.1	3.1	1.3	2.7	1.2
	11.5	5.0	5.0	3.8	5.7	2.3	5.0	2.2
	23.0	9.0	9.0	7.5	10.0	4.3	10.0	4.4
Diesel tubwell	1.2	0.6	0.6	0.4	0.7	0.2	0.6	0.2
	2.2	1.0	1.0	0.7	1.1	0.4	1.0	0.4
	2.9	1.2	1.2	0.9	1.9	0.5	1.2	0.5
	6.5	1.7	1.7	1.4	2.9	1.0	1.7	0.8
	11.5	2.3	2.3	1.9	4.0	1.5	2.4	1.3
	23.0	2.8	2.8	2.5	5.0	2.5	3.1	1.8

For large farmers, who can run the tubewells very near the rated capacity, electric tubewell is highly profitable, with or without access to cheap source of finance from institutional sources. The introduction of fixed power tariff tends to widen the gap between returns from an electric tubewells and a diesel tubewell as a tubewell approaches its rated capacity. Small farmers having access to institutional funds are in a position to benefit from tubewell technology only if they can surmount their farm size disability through mutual co-operation in utilising the tubewell investment. Otherwise, they would find traditional well irrigation more beneficial. But without access to institutional funds, many of them would find even this option uneconomic. In point of fact the B-C ratios for small farmers with traditional wells may turn out to be much less than the levels indicated if they experience external diseconomies of the working of tubewells located near their wells.

III

The official statistics of groundwater structures suffer from substantial upward bias. In view of this one feels apprehensive that the Groundwater Directorate's estimates about groundwater draft may also be biased upwards. According to these estimates, the groundwater draft during 1975-76 amounted to 40 per cent of the groundwater potential of east UP. Though this level of utilisation is indicative of serious underexploitation of groundwater resources of the region under study, still there are reasons to believe it is on the high side, possibly by 10 percentage points. A disaggregative analysis reveals that four districts have rather high rates of groundwater exploitation. These are Faizabad with 81 per cent, Jaunpur with 69 per cent, Varanasi with 62 per cent and Allahabad with 61 per cent. Their levels of utilisation are not only above the regional level or state average of 48 per cent, but are also in excess of the average level of 60 per cent for west UP region.

From the viewpoint of analysing factors behind inter-district variations in the level of utilisation of groundwater potential, the estimates of the Groundwater Directorate are particularly unsuited because of the adoption of uniform norms of annual groundwater draft per well or tubewell for all districts of the UP plains. Irrespective of the magnitude of these norms, their uniform adoption for all districts eliminates considerable amount of variance in district-wise estimates of groundwater utilisation that arises because of wide inter-district variations in intensity of irrigation, crop pattern, rainfall, etc.

Thus, both because of a serious upward bias in the stock of private groundwater structures and the application of uniform norms of groundwater draft per structure, the district-wise estimates of groundwater withdrawals of the Groundwater Directorate do not appropriately convey the inter-district differentials in the level of groundwater extractions within the UP plains. With a view to having a more accurate picture of inter-district variations, a new methodology has been evolved, whereby the data base has been shifted from the stock of groundwater structures to the area irrigated by groundwater sources, and the norm for converting an acre irrigated by groundwater sources is objectively varied with a district.

It may be useful to distinguish between (i) actual or observed volume of groundwater draft ('V') in a year and (ii) potential volume of groundwater draft ('C') possible in a year from the combined installed capacity of wells/tubewells. Evidently, V and C are equal when the installed capacity of wells/tubewells is fully utilised in a year. If 'P' is the groundwater potential, both V/P and C/P are meaningful measures of the level utilisation of groundwater potential. Though the measure C/P is extremely useful while planning fresh installations of wells and tubewells in an area, for a variety of reasons our primary concern is with the other measure V/P. To begin with, it is more comprehensive of the two because it also takes into account the level of capacity utilisation of wells/tubewells. In fact, it is nothing but an exact product of the level of capacity utilisation of wells/tubewells and the other measure.

$$\frac{V}{P} = \frac{V}{C} \times \frac{C}{P}$$

Secondly, assessment of installed tapping capacity C is itself no easy task once one begins probing in right earnest the question of installed capacity of a well or tubewell. Even if this task is resolved through pragmatism, the district-wise estimates of C would suffer from serious upward distortion if official district figures of masonary wells, pumpsets, etc as reported by the state's minor irrigation department are to be used.

The principal factors taken into account for varying the norm are inter-district differentials in respect to intensity of irrigation, crop pattern of irrigated agriculture, rainfall and the source-wise composition of groundwater-irrigated area. Admittedly, some other important factors, such as soil and natural sub-soil irrigation in Terai lands, have been left out of account. In view of this and some other limitations an attempt has been made to offer two sets of estimates of per cent utilisation of groundwater potential, namely, U_1 and U_2, with $U_2 > U_1$. The average of U_2 and U_1 is being taken as the most probable estimated value of per cent utilisation of groundwater potential of a district (Table 10.8).

TABLE 10. 8

Two Alternative Estimates of Groundwater Utilisation, 1970-71.

District/ Region	*Alternative Estimate of Groundwater Extraction ('000 hectare-metres)*		*Groundwater Potential ('000 hectare-metres)*	*Percentage Exploitation of Groundwater Potential*	
	V_1	V_2	P	U_1 $\frac{\text{Col 2}}{\text{Col 4}} \times 100$	U_2 $\frac{\text{Col 3}}{\text{Col 4}} \times 100$
(1)	*(2)*	*(3)*	*(4)*	*(5)*	*(6)*
Agra	40	36	90	44	40
Aligarh	89	82	163	55	50
Bijnor	34	51	164	21	31
Badaun	40	46	124	32	37
Bareilly	10	15	95	11	16
Bullandshahr	105	98	174	60	56
Etah	39	38	134	29	28
Etawah	20	21	100	20	21
Farrukhabad	33	36	105	31	34
Mainpuri	38	38	146	26	26
Mathura	39	32	141	28	23
Meerut	123	123	204	60	60
Moradabad	67	89	138	49	64
Muzaffarnagar	69	73	152	45	48
Rampur	7	10	75	9	13
Pilibhit	6	10	74	8	14
Saharanpur	44	58	176	25	33
Shahjahanpur	17	24	182	9	13
West UP	**820**	**880**	**2437**	**34**	**36**
Allahabad	17	23	126	13	18

cont.

Table 10.8 cont.

(1)	(2)	(3)	(4)	(5)	(6)
Pratapgarh	10	13	107	9	12
Jaunpur	31	43	116	27	37
Ghazipur	21	31	141	15	22
Ballia	15	21	159	9	13
Gorakhpur	22	42	183	12	23
Deoria	41	65	110	37	59
Basti	32	56	247	13	23
Azamgarh	46	65	204	23	32
Faizabad	28	39	124	23	31
Gonda	22	35	225	10	16
Bahraich	6	10	98	6	10
Sultanpur	16	22	133	12	17
Varanasi	22	32	105	21	30
Mirzapur	3	5	552	6	10
East UP	**332**	**502**	**2133**	**16**	**24**
Bara Banki	12	17	97	12	18
Fatehpur	12	15	113	11	13
Hardoi	11	13	151	7	9
Kanpur	16	18	139	12	13
Kheri	10	15	90	11	17
Lucknow	4	5	57	7	9
Rai Bareili	10	13	134	7	10
Sitapur	7	9	125	6	7
Unnao	4	5	104	4	5
Central UP	**86**	**110**	**1010**	**9**	**11**
UP Plains	**1238**	**1492**	**5580**	**22**	**27**

The latest year for which the author could operationalise his methodology is 1970-71, for which district-wise statistics based on the first census of agriculture are readily available. The earliest year for which the estimates have been obtained is 1950-51. It is found that there was a rise of about 5 percentage points in per cent utilisation of groundwater potential in east UP during the two decade period 1950-51 to 1970-71. The level of about 20 per cent ((U_1 = 16 and U_2 = 24) attained in 1970-71 was below that of west UP (34-36 per cent) but above that of central UP (9-11 per cent), whereas the average for the UP plains as a whole stood in the range 22-27 per cent. Within the east UP region, Deoria, Jaunpur, Faizabad, Azamgarh and

Varanasi districts were above the regional average, but much behind the levels attained in Meerut, Bullandshahr, Muzaffarnagar and Aligarh districts of west UP.

FACTORS BEHIND LOW UTILISATION

Natural Factors : Soil and climate endowment of east UP being more favourable, the need for artificial irrigation is less intense than in the north-western belt. Though precise figures are lacking, there is no doubt that the proportion of water-retentive soils is much higher in east UP than in the largely semi-arid north-western belt. Its higher rainfall diminishes the need for irrigation during the Kharif season when the flat topography would pose a problem of drainage if irrigation is introduced without field drains. The greater flood-proneness of the region exposes fixed investments in groundwater to damage from floods, particularly in the flood plains which are said to account for half the geographical area of east UP. Thus an average farmer in east UP has to contend with greater risk factor whereas his risk-bearing capacity is lower because of the smaller base of his assets, as indicated by smaller size of his land holding.

Infrastructural Factors : East UP is comparatively disadvantaged in the matter of infrastructural facilities which have a close bearing on the private benefit-cost ratio of private investments in groundwater irrigation. More specifically, agricultural extension services, short-term credit for crop production, road transportation to the market centres and rural electrification are still lagging behind the levels encountered in north-western India. More long-term institutional credit has been available since the early seventies. In fact, the reason for the marginal rise in the level of groundwater utilisation during the first twenty years of planning can be squarely attributed to the slow development of infrastructural facilities conducive to private investments in groundwater irrigation. The advent of HYV seeds in the mid-sixties made investments quite remunerative but such investments took place slowly because supporting infrastructural facilities for agriculture were lacking.

Socio-political Factors : Among these we may note the inhibitive role of fragmentation of holdings, widespread tenancy, and a generally fatalistic attitude towards life. These factors do change over time, but rather slowly.

CONCLUDING REMARKS

All over the world, overexploitation of groundwater resources has been reported from many tracts where such resources could be tapped by tubewell technology. The untoward consequences of such 'mining' have been permanent lowering of water table, or damage to groundwater resources from the intrusion of saline water, or both. Such adverse developments have been particularly noticed in semi-arid tracts characterised by low rainfall; intense demand for irrigation due to low rainfall, sandy soils, and arid climate; absence of perennial rivers; and existence of saline/brackish waters in the neighbourhood of groundwater resources.

No such adverse developments need be feared in the case of east UP groundwater resources. To begin with, the region is far away from the point of 'mining'. In the more recent years of increase in the number of tubewells, its level of utilisation rose by 10 percentage points. Even if this spurt continues unabated in future, another 30 years are required to reach the mark of 100 per cent utilisation of groundwater potential. Thus expansion of groundwater irrigation can go on unimpeded till the turn of this century. All that is needed is regulation of spatial location of wells so as to eliminate mutual interference whereby each well owner imposes external diseconomies of his neighbours. Such short-run external diseconomies -- to be disinguished from long-run external diseconomies created by permanent recession in water table – have been reported by the owners of traditional wells whose water yields as well as operating costs are adversely affected by tubewells in their immediate vicinity. This externality could be an important reason why dug wells have fallen out of favour with small and medium farmers of the region.

It is not known what precisely have been the external diseconomies of the recent spurt in private tubewell irrigation in east UP. So long as traditional and modern modes of groundwater irrigation co-exist on an extensive scale, it is difficult to visualise any ideal scheme of regulation which avoids the imposition of externality of private tubewell irrigation on dug well irrigation. Short of forbidding the installation of private tubewells which are now far more economical than dug wells for large and medium farmers of east

UP; the state can do nothing more than strive for regulating the mutual interference among tubewells in a given zone.

At first thought, it looks an attractive idea to protect the interest of small and marginal farmers through public irrigation. Since canal irrigation works much more smoothly than public tubewells, it needs priority in expansion. Their supplies can be strengthened with large size augmentation tubewells which can be supplied with power through separate transmission lines that would maintain continuous electricity supply to the tubewells even under conditions of overall power shortages. Public tubewells with independent commands may prove a far less attractive option to the small and marginal farmers as well as the state. These are more difficult to maintain in working order than augmentation tubewells, with the result that water supplies from them are unreliable for the beneficiaries. Their unit cost of irrigation may become unduly high if their annual running hours are below the optimum level, either because of bottlenecks on the supply side or low demand because of existence of private tubewells in their commands, sale of surplus water by private tubewell owners to other farmers, less intensive agriculture being practised, etc.

In contrast to public irrigation, the state can encourage small and marginal farmers to exploit tubewell technology co-operatively. However, the success of such as scheme hinges on the existence of spirit of co-operation among the beneficiaries and on their land-holding being in close proxmity to one another. These are rather stringent requirements because of which few cooperative tubewells have come into being in east UP. Perhaps, subsidy in the shape of low interest rate on term loans from institutional agencies help small and marginal farmers to perceive better their mutual advantage in installing a co-operative tubewell.

Availability of institutional funds makes the smallest size tubewell a worthwhile investment proposition for a 5-acre farmer who is hardly likely to use one-third of the irrigation capacity of such a tubewell. In fact, if small farmers can indirectly overcome their size disability with regard to a tubewell, they would find tubewell much more rewarding than the alternative mode of dug-cum-bore well operated by animal-operated or power-driven water-lifts. If they can thus utilise fully the irrigation capacity of a tubewell, their benefit-cost ratio from an electric tubewell irrigation could be somewhere between 4 and 10 -- the higher ratio prevails when access to institutional sources of finance exists. Were they to exploit fully the capacity of dug-cum-bore wells on their individual farms, their B-C ratio might be 2.7 at the most, and that too if (a) they put a very low valuation on their own labour used in the construction and maintenance of such wells, and (b) they enjoy access to institutional finance. In case they finance their investments at high rates prevailing in the unorganised capital market of the area, the ratio would then be 1.7 at the most. These ratios are for animal operated dug-cum-bore wells. Power-driven pumpsets barely appear economic for small and medium farmers even when these are financed from institutional funds.

Inter-district differential in the level of exploitation of groundwater potential within the plains of Uttar Pradesh are also correlated with the level of rural electrification. However, since rural electrification is found to be correlated with other variables (notably road transport) which in turn are positively correlated with the per cent utilisation of groundwater potential, the correlation between it and per cent utilisation of groundwater potential gets exaggerated in view of this multi-collinearity. All one can conclude is that maximal impact of rural electrification in promoting groundwater utilisation would occur in those tracts as are concomitantly having infrastructural facilities and institutional reforms. It is because of comparatively greater concomitance or agglomeration of factors promoting intensive agriculture that relationship between rural electrification and per cent utilisation of groundwater potential is stronger in west UP than in east UP. Among infrastructural facilities, the special need of east UP now is with regard to flood protection and drainage. Impounding monsoon torrents in the Nepal hills is the best protection against floods. As for drainage, investment in cross-drainage needs to be taken care of while expanding the network of roads, railways and canals in this flat region. Furthermore, returns from kharif irrigation can be raised substantially if farmers are encouraged to invest in field drainage.

APPENDIX *

ESTIMATION OF GROUNDWATER WITHDRAWALS

Groundwater draft can be assessed in two ways. One is the stock or unit approach in which the estimation is done with the help of the stock of groundwater structures, namely, wells and tubewells. The other is the area approach, in which the estimation is done with the help of the statistics of area irrigated by groundwater sources. While the norms of groundwater withdrawal per groundwater structure are required in the first approach, one requires a norm of irrigation delta (height or depth of irrigation application) for the second approach. For ascertaining district-wise estimates of groundwater withdrawals in a year, the main problem is one of varying these norms according to the characteristics of a district. How this variation has been achieved in the case under study is outlined in what follows.

For Meerut district, the Roorkee Division of the Groundwater Investigation Organisation, U.P. did a detailed study of groundwater potential and groundwater draft during 1970-71.[1] Dividing this draft by the corresponding area irrigated by groundwater sources, we get the irrigation delta (H) for Meerut district. Evidently, this delta is postulated to vary over the districts of the Gangetic plains of U.P. in the following manner:

a) soil-weather considerations: 'H' is inversely related to rainfall and clayeyness (moisture-retentive ability) of soil of a district.
b) intensity of irrigation: Defining it as 'gross' irrigated acreage divided by 'net' irrigated acreage, 'H' would vary positively with intensity of irrigation to groundwater irrigation. In this study, multiple irrigated cropping and intensity of irrigation are used synonymously.

---

* not published in *EPW*.

1. Source: *Technical Feasibility of Groundwater Development on Compact Area Basis: Meerut District (mimeo)*. According to this report, the net area irrigated by groundwater and consequent withdrawal of groundwater in 1970-71 in Meerut district were 216, 496 hectares and 1371 million cubic metres (137,100 hectare-metres), respectively.

c) crop mix: 'H' would vary positively with the importance of crops requiring heavy irrigation (e.g. HYV crops, sugarcane, potato, banana, paddy, etc.) in district crop pattern on lands irrigated by groundwater sources.
d) composition of groundwater sources: Even if *ceteris paribus* situation about (a), (b) and (c) factors prevails, 'H' is known to be higher for private tubewell irrigation than for open or dug well irrigation.

Given H_m (m for Meerut district), one may write H_j for j th district of U.P., using the principle of proportionality, in the following manner:

$$H_j = H_m . \frac{(I_j).}{(I_m)} \frac{(\bar{A}_j . A_m).}{(A_m . A_j)} \frac{(R_m)}{(R_j)} \qquad ......(1)$$

where : H is average groundwater delta per net hectare of area irrigated by groundwater sources;

'I' is intensity of groundwater irrigated cropping adjusted for crop pattern (see below)
'A' is net groundwater-irrigated area duly adjusted for groundwater source mix only (see below)
'R' is normal rainfall of a district
j and m stand for j th and Meerut districts.

The adjustment in 'I' for crop pattern is done by weighting irrigated area under a crop by its irrigation water requirements. Thus, instead of the conventional measure $(TG_i) \div A$ the following measure of 'I' is used:

$$I = (T h_i . G_i) \div A$$

where :

'G_i' is the irrigated area under i th crop;
'A' is total net irrigated area corresponding to total gross irrigated area.
'h_i' is relative water requirements of i th crop with requirements of wheat = 1

Likewise, the adjustment in net irrigated area 'A' is done as follows:

$$\bar{A} = \Sigma\, a_i . W_i$$

where 'ai' is the net area irrigated by i th source of groundwater irrigation in a district, with $\Sigma\, a_i = A$

'W_i' is weight attached to i th source of groundwater irrigation with $W_i = 1$ for private tubewell irrigated area.

Before spelling out how the formula (1) has been operationalised in this study it is necessary to make a few observations about this formulation. First of all, it does not take care of inter-district variations in respect of water-retentive nature of soil; natural sub-soil irrigation (e.g. in 'Tarai' lands); level of agricultural development, especially in respect of HYV seeds and fertilizer use; and the extent to which groundwater irrigation is supplementary with other irrigation in a district. Secondly, the above formulation rests on the assumption of proportionality. That is, 'H' is assumed to vary in direct proportion to 'I' and 'A', and in inverse proportion to 'R'. While one feels the need for assuming non-linear relationships here, one does not have the necessary data for departing from the world of linearity. Thirdly, since the effect of intensity of irrigation and crop pattern are taken care of in the measure 'I', the relative weights 'W's. in 'A' are supposed to reflect only the pure effect of mix of groundwater sources.

While using the formula (1) for estimating groundwater irrigation delta for each district of U.P., the following assumptions have been made:

(i) In the absence of information about crop-wise area irrigated by groundwater sources, 'I' has been computed from the overall irrigated crop pattern and total net irrigated area of each district. This tantamounts to an assumption of equality between intensity of groundwater-irrigated cropping and intensity of surface-irrigated cropping in a district.

(ii) Adjustment for crop pattern is partial in the computation of 'I', as relative irrigation needs of wheat and sugarcane only have been taken into account. However, wheat and sugarcane account for a substantial share of irrigated crops in every district of U.P. plains. 'h' for sugarcane is taken 2.5

(iii) Crop for crop, the water application differential for an acre of groundwater irrigated land is assumed to be as follows:

	W_i
Private tubewell-irrigated	1.0
Government tubewell-irrigated	0.8
Dug well-irrigated	0.7

Besides data about net area irrigated by wells and tubewells, we have reliable information about the number of government tubewells. By making some assumption with regard to net area irrigated per government tubewell[2], one thus completes the following information about groundwater mix for every district:

(1) net area irrigated by government tubewells;
(2) net area irrigated by private tubewells;
(3) net area irrigated by open wells.

We had the following additional information to help us in our estimational work:

.... comparative irrigation of wheat and sugarcane crops in Meerut and Faizabad Divisions for the year 1962-63.[3]

2. The government tubewells, according to the Agricultural Census of Uttar Pradesh, irrigated (net) 838 thousand hectares in the whole of U.P. in 1970-71. Since there were about 10,000 government tubewells in operation in 1970-71, net area irrigated per tubewell comes to about 83 hectares. In inferring district-wise net area irrigated by government tubewells we have multiplied the number of government tubewells in 1970-71 in each district by a factor of 80.

3. Source: Shridhar Misra, *A Comparative Study of the Economics of Minor Sources of Irrigation in Uttar Pradesh,* Oxford and IBH Publishing Co., Calcutta, 1968. Irrigation is measured in terms of 'number of irrigations', and not volume of irrigation. The primary information is reproduced below:

Irrigation Source	No. of irrigations	
I. *Meerut Division*	sugarcane	wheat
1. persian wheel	5.74	2.77
2. state tubewell	5.33	2.27
3. private tubewell	6.38	2.99
II. *Faizabad Division*		
4. ordinary masonry well	2.78	1.12
5. state tubewell	3.54	1.85

...comparative irrigation by different groundwater sources for Aligarh district for 1966-67[4] and Meerut district, 1968-69[5].

Two sets of district-wise estimates of groundwater withdrawal have been obtained - one with the formula (1) and one with modified formula (1). The modification is done by dropping rainfall from this formula. Thus, two estimates of withdrawal, V_1 and V_2, are obtained, which in turn yield two estimates of percent utilisation of groundwater potential (P), namely, U_1 and U_2. These estimates are listed in Table 10.8.

While knowledge about 'H' for at least one district is required for operationalizing this methodology of varying norms, it may be pointed out analysis of inter-district variations of 'V' as well as 'U' is invariant with the magnitude of 'H'. In other words, errors in selecting 'H' do not affect the inter-district correlation analysis.

---

4. Source: T.V. Moorti, *A Comparative Study of Well Irrigation in Aligarh District*, India, Occasional paper no. 29, Department of Agricultural Economics, Cornell University, 1970.
For four kinds of groundwater appliances we find the following comparative picture of irrigation (in terms of millimeters of water per hectare):

	State tubewell	Private tubewell	Persian wheel	Charsa
1. Simple (unweighted) average for four crops common to all the groundwater sources.-desi bajra, desi wheat, wheatgram mix, and barley & peas mix	266 (0.80)	333 (1.00)	245 (0.74)	209 (0.63)
2. Simple average for 8 crops* common to all sources but charsa	285 (0.80)	357 (1.00)	266 (0.75)	-
3. Simple average for all 14 crops **	285 (0.79)	362 (1.00)	261 (0.72)	209 (0.58)

* Besides the 4 common crops, others are : Mexican wheat, desi maize, peas and barley and peas mix.

** There is no sugarcane or paddy in this set.

Note : Figures in brackets are relative irrigation, with private tubewell irrigation = 1.00.

5. Source: S.M. Patel and K.V. Patel, *Economics of Tubewell Irrigation*, Indian Institute of Management, Ahmedabad, 1971. The comparative irrigation picture for Meerut district for 1968-69 was as follows:

	Sugarcane	*Wheat*
1. Irrigation from state tubewell	81 inches* (3.24)	25* inches

cont.

fn 5 contd.

2. Irrigation from private electric tubewell	45 (2.37)	19
3. Irrigation from private diesel tubewell	41 (2.05)	20
4. Irrigation from wells with diesel pumpsets	31 (2.84)	11

Note: Figures in brackets are relative irrigation of sugarcane with wheat = 1.

* Estimates derived by the authors indirectly by dividing reported irrigation expenditure by Rs. two, the tubewell irrigation rate in 1968.

11

Demand for Irrigation
(A Case Study of Government Tubewells in Uttar Pradesh*)

SCOPE OF THE STUDY

The purpose in this short paper is to present some estimates of elasticity of demand for irrigation water with respect to four variables, namely, (1) price of farm output, (2) price of irrigation water, (3) normal rainfall, and (4) deviation from normal rainfall. A single-equation linear model has been employed to the time-series data in respect of State tubewells in Uttar Pradesh. The study provides a basis for important policy implications which are spelt out towards the end of the paper.

THE DATA

The basic data about the level of utilization of tubewell are obtained from the Office of the Chief Engineer, Irrigation Department, Uttar Pradesh. Another useful source has been the administration report of the Irrigation Department, U.P.[1] Instead of using the all-India statistics of wholesale prices of agricultural commodities, the index numbers of wholesale prices of agricultural commodities compiled by the Office of the Director of Economics and Statistics, Uttar Pradesh, have been used to represent farm prices received by U.P. farmers drawing water from State tubewells. Other sources of data are listed at appropriate

---

* Reprinted *Indian Journal of Agricultural Economics* Vol. XXVIII No.2 April-June 1973.

1. Published annually under the title Irrigation Administration Report of Uttar Pradesh by the Superintendent, Printing and Stationery, Uttar Pradesh, Allahabad.

places in the text. The measurement of the irrigation rate for tubewell water needs three observations. First, it is an effective rate, in the sense that any rebate or surcharge on basic rate is fully taken into account. Second, only the basic rates for electricity-driven tubewells could be considered. Third, till September, 1967, tubewell water in U.P. was sold under one-part tariff based on volume of water alone. Thereafter, a two-part but optional tariff has come into force. This, however, does not affect our results because our time-series analysis is confined to 1946-67 during which period strictly volumetric basis of rate fixation prevailed. In point of fact, it is this volumetric basis of irrigation rates, unlike the crop basis of rates prevailing in most major and minor irrigation works, that makes it possible to undertake a study concerning demand for irrigation water.

The measure of water consumption in this study is the average annual running hours per tubewell. It has been possible to use running hours as a proxy for tubewell water consumption because of the fact that all the tubewells, in terms of their water discharge rate, are of the same size, namely, 1.5 cusecs or 30,000 gallons per hour. We believe that this is decisively a better measure of demand for irrigation water than the conventional measure of acreage irrigated, whereby the intensity of irrigation is ignored.

Finally, a few words may be said about tubewells being in development phase. A newly commissioned tubewell, from which water is being released for irrigation, according to the State Irrigation Department, is deemed to be in development phase for three years after the date of its commissioning. This development stage is essentially a formative stage in the sense that any one or more of the following situations may prevail since the commissioning of a tubewell.

1. The full complement of its water courses for the entire command area is not constructed due to difficulty in acquiring land for the main water courses.
2. The water courses, though complete, are not fully lined because of (a) shortage of lining materials and (b) lack of synchronization in constructing water courses and the central installation of a tubewell.
3. The farmers have not fully constructed their own field channels for some reasons.

THE MODEL

A linear multiple regression model of the following type has been used:

$$Y = a + \sum_{i=1}^{5} b_i X_i$$

where Y = annual running hours per tubewell in operation in mid-year.

X_i = 'i' th explanatory variable.
b_i = the regression coefficient of X_i.
a = constant.

The model has been estimated by the method of least squares. Being a single-equation model, the estimates do suffer from the single-equation bias of an unknown magnitude. The expected signs of the regression parameters are that all b_i's other than b_2 are negative if X_i's are defined as follows:

X_1 = Index of actual rainfall with normal rainfall (X_3) equal to 100

or

actual rainfall in millimetres in the case of Etah district study.

X_2 = Index number of agricultural wholesale prices with 1953=100.

X_3 = The weighted average of districtwise normal rainfall in millimetres; the weights beings proportional to the number of tubewells in a district.

X_4 = Effective irrigation rate in nPs per 10,000 gallons of water.

X_5 = The percentage of tubewells in development stage.

Three observations about the variables X_2, X_3 and X_5 may be in order. First, the index number of agricultural prices (X_2) is available on a calendar year basis, whereas the explained variable (Y) is on a financial year basis. An attempt is made in this paper to introduce a dynamic element in the model by bringing about a 3-month lag between Y and X_2 as follows. The value of Y in a year, say 1960-61, is regressed on the value of X_2 in 1960. In doing so, we have departed from the conventional practice of lagging X_2 by a full crop year in studies relating to farmers' response to prices. The rationale for doing so is spelt out in the following para.

The allocative decisions may be distinguished in the following way : (1) decision with regard to area to be allocated to a crop from his given holding and (2) decision with regard to the optimum quantity of a variable input for his given crop acreage. For a decision of type (1), the usual lag of one crop year has been found to be a good proxy for farmers' expectations in respect of crop prices in several studies in the literature. For a decision of type (2), the conventional procedure is questionable. The price of a crop ruling over the growth period of the crop has a definite bearing in determining a farmer's decision on buying the quantities of the variable inputs for which he pays on a *pro rata basis.* Tubewell water is purchased variable input whose intensification would be partly determined by the course of agricultural prices since the date of sowing (or decision to sow) of a crop. For, he is buying and applying this input not in one lot at the time of sowing but in numerous lots over the entire gestation period of a crop. For instance, the decision to buy the quantum of irrigation water towards the maturing stage of a crop is likely to be influenced *inter alia* by crop prices of the immediate past rather than of a year ago.

The second observation pertains to the introduction of the variable normal rainfall (X_3). There has been a perceptible shift in the regional dispersal of State tubewells in U.P. Back in the 'thirties the State Government started installing tubewells in the comparatively low rainfall districts of West U.P. By the beginning of the planning era in 1951-52, 96 per cent of the tubewells were in West U.P. The 'fifties witnessed a major locational shift in installing new tubewells. Of the 4,000 tubewells so added, three-fourths went to Eastern and Central parts of U.P. which experience more rainfall than the western parts of the State. The irrigation engineers (including the administrators) have been trying to explain away the falling trend in the aggregate utilization of State tubewells in terms of the locational shift, with the tacit assumption : the higher the normal rainfall, the lower the expected utilization of an irrigation work like State tubewells where the rate is tied up, unlike in canal irrigation, essentially with the quantum of water purchased. The importance of the argument of locational shift is being tested here in the model. The third noteworthy observation pertains to the variable X_5. The rationale for introducing this variable, which partakes the character of a variable partly on the demand side and partly on the supply side, is that the irrigation engineers and administrators of the State Irrigation Department strongly believe that the downward trend in the level of utilization of State tubewells is, among other things, due to the increasing proportion of new tubewells being in a development phase.

Before closing this section, it may by pointed out that the model has two noteworthy specification errors and one noteworthy error of observation in variable X_2. The specification error emanates from the absence of two explanatory variables that could capture the intra-year variations in rainfall and the crop pattern. *A priori,* it is difficult to say how the absence of these variables affects the estimates of the regression parameters. The error of observation in X_2 arises due to the fact that temporal movements in farm prices received by farmers may not be fully captured by the index of wholesale prices of agricultural commodities.

Both the specification biases could not be rectified because of paucity of readily available data on the two missing explanatory variables and the index of farm prices for U.P. for the period under study. However, our hunch is that these biases are of tolerable magnitudes.

REGRESSION RESULTS

The linear multiple regression equation derived from the basic data in Table 11.1 is as follows :

$$\underset{}{Y = 14366} - \underset{(2.67)^*}{13.18X_1} + \underset{(2.35)}{9.62X_2} - \underset{(1.63)}{12.63X_3} - \underset{(5.33)}{2.26X_4} + \underset{(6.27)}{5.58X_5} \quad ....(1)$$

$$R^2 = 0.97$$

(* Standard errors of regression coefficients.)

Contrary to a *priori* expectations, the regression coefficient of tubewells in development phase (X_5) has a positive sign. However, it need not cause any concern because it is insignificant, with a low t-value of 0.9. What is really noteworthy is the insignificance of the coefficient of irrigation rate for tubewell water (X_4), as its standard error is more than double its own value. All the other three explanatory variables are highly significant at 1 per cent level of significance.

TABLE 11.1
Basic Data for Time-Series Analysis of the State Tubewells : 1946-47 to 1966-67

Year	*Y*	X_1	X_2	X_3	X_4	X_5
1946-47	3,402	112	72	826	29	14
1947-48	3,355	97	84	826	29	17
1948-49	3,051	128	105	826	29	14
1949-50	3,671	101	124	826	29	12
1950-51	3,536	112	103	826	29	11
1951-52	3,805	90	106	826	44	10
1952-53	3,650	86	103	839	44	12
1953-54	2,967	105	100	864	63	18
1954-55	2,659	88	83	890	63	29
1955-56	2,184	136	68	908	50	38
1956-57	1,751	132	86	925	50	40
1957-58	2,234	92	95	941	50	37
1958-59	1,937	131	108	939	50	31
1959-60	2,306	85	106	940	50	25
1960-61	2,005	114	101	939	50	19
1961-62	1,623	128	99	938	63	14
1962-63	2,151	90	104	940	63	12
1963-64	2,124	108	112	937	63	14
1964-65	2,389	103	150	933	63	14
1965-66	3,088	75	164	935	63	13
1966-67	3,369	79	169	938	63	11

Notes : (1) Y stands for annual running hours per tubewell in operation in mid-year.
(2) X_1 stands for the index of actual rainfall with normal rainfall taken as 100.
(3) X_2 stands for the index number of agricultural wholesale prices, with 1953 = 100.
(4) X_3 stands for the weighted average of districtwise normal rainfall in millimetres, the weights being proportional to the number of tubewells in a district.
(5) X_4 stands for irrigation rate for tubewell water in nPs per 10,000 gallons.
(6) X_5 stands for percentage of tubewells in development stage.

Deleting X_4 and X_5 variables from the regression, the equation (1) assumes the following shape :

$$Y = 14298 - \underset{(2.49)^*}{12.55\,X_1} + \underset{(1.85)}{8.34\,X_2} - \underset{(0.83)}{12.48\,X_3} \quad \dots\dots \quad (2)$$

$$R^2 = 0.9461$$

$$\bar{R}^2 = 0.9334$$

(*Standard errors of regression coefficients.)

Normal rainfall variable X_3 carries the largest explanatory power, with a value of --0.96 for its partial correlation coefficient with Y. One possible reason for this high explanatory power of X_3 is the relationship (r = 0.78) it has with X_4. In point of fact, the statistical insignificance of X_4, noted earlier, could as well be attributed to this problem of multi-collinearity. In order to have a firmer idea about the role of X_4 in influencing the farmers' demand for water, it is desirable to look into another set of data that is free from this problem of multi-collinearity. The following analysis based on the Etah district data satisfies this statistical criterion. The data in respect of Etah district analysis are set out in Table 11.2. The data for the ten tubewells are culled from a mimeographed study entitled "Economic Study of State and Private Tubewells in Varanasi and Etah Districts of U.P." prepared by the Planning Research and Action Institute, Lucknow, in May, 1968.

The level of utilization of the ten Etah tubewells was quite high in 1952-53 as they had been constructed mainly during the first decade of tubewell development in western U.P.[2] However, the changing pattern of utilization of these tubewells between 1952-53 and 1966-67 is broadly similar to the one found in the case of all the State tubewells taken together. The utilization tends to fall sharply during the 'fifties and then rise during the sixties'.

The precise influence of actual rainfall (X_1), index number of agricultural prices (X_2) and price of tubewell water (X_4) on the dependent variable Y is as follows :

$$Y = 3293 - \underset{(0.82)}{2.33\,X_1} + \underset{(6.23)}{19.97\,X_2} + \underset{(21.74)}{13.07\,X_4} \quad \dots\dots \quad (3)$$

2. The programme of installing State tubewells began in U.P. in the mid-thirties.

TABLE 11.2
Basic Data about Ten Tubewells in Etah District

Year	*Average running hours per tube-well* (Y)	*Rainfall (millimetres)* (X_1)	*Index number of agricultural wholesale prices* (X_2)	*Irrigation rate per 10,000 gallons* (X_4) *nPs*
1952-53	4,286	596	103	39*
1953-54	3,430	657	100	63
1954-55	2,919	611	83	63
1955-56	2,519	837	68	50
1956-57	1,782	897	86	50
1957-58	2,130	621	95	50
1958-59	2,010	1,076	108	50
1959-60	3,075	572	106	50
1960-61	1,972	1,168	101	50
1961-62	2,119	945	99	63
1962-63	2,813	533	104	63
1963-64	3,359	828	112	63
1964-65	3,842	695	150	63
1965-66	3,953	544	164	63
1966-67	5,128	539	169	63

* In view of the fact that cane is an unimportant crop in the irrigated agriculture of Etah district, this rate refers to the effective irrigation price for crops other than sugarcane.

The R^2 of equation (3) is about 0.7223, which diminishes to about 0.6214 when adjusted for the 11 degrees of freedom of the equation. This not too high value of the coefficient of determination indicates that some relevant variables, like crop pattern, are missing. Keeping in mind this limitation on the equation, we may proceed ahead with the interpretation of the results of this regression analysis.

All the three explanatory variables in equation (3) have the right expected signs. While rainfall (X_1) and agricultural prices (X_2) turn out to be with significant regression coefficients (at 5 per cent level of significance), the regression coefficient of tubewell water rate (X_4) does not become significant even when the level of significance is lowered to 50 per cent.[3] In the present case, the insignificance cannot be attributed to any phenomenon of multi-collinearity. For, the coefficients of correlation of X_4 with X_1 as well as X_2 are small in magnitude, as can be seen from Table 11.3.

TABLE 11.3
Correlation Matrix for Data in Respect of Ten Tubewells of Etah District

X_1	X_2	X_4	Y
1.0000			
--.3569	1.0000		
--.2569	.4330	1.0000	
--.6720	.7176	.2741	1.0000

Note : X_1 stands for rainfall; X_2 stands for agricultural prices; X_4 stands for irrigation rate, and Y stands for running hours per tubewell.

The value of the elasticity of irrigation demand with respect to X_1, X_2 and X_4, at the mean levels of these variables, are (--) 0.57, 0.72 and (--) 0.24 respectively.[4] What is striking is that the demand is quite elastic with respect to agricultural prices, and not with the price of tubewell water.

The values of elasticity coefficients emerging from the Etah district data are certainly not representative of the better part of Uttar Pradesh. Because of low natural precipitation (about 700 mm.) the agriculture in Etah is distinguished by a crop pattern that has two distinct features: (i) emphasis on millets and (ii) little share of commercial crops, with sugarcane accounting for a negligible proportion in the whole. These distinctive characteristics are representative of the nine

3. Dropping X_4 from the equation (3), we get the following equation:
$$Y = 2669 - \underset{(0.79)}{2.27}X_1 + \underset{(5.61)}{18.55}X_2,\ R^2 = 0.732,\ \bar{R}^2 = 0.6415$$
4. These may be compared with the values of elasticity coefficients of (--) 0.48 for X_1, 0.33 for X_2 and (--) 0.41 for X_3 that emerge from the equation (2).

districts of South-Western U.P.[5] Therefore, the above magnitudes of elasticity are relevant for this region of U.P.

IMPLICATIONS

(1) The upshot of the time-series analysis is that the level of utilization of the tubewells is more sensitive to the state of agricultural prices rather than the price of tubewell water. This will suggest that the phenomenon of under-utilization of the newly-created irrigation potential that was noticed in the 'fifties could be attributed to the depressed price level of agricultural commodities during 1953-56 rather than to the absence of right promotional price policy for irrigation water.

(2) There is a widespread feeling that irrigation water is seriously under-priced in India, with the result that the Government incurs substantial losses in running the irrigation projects. State tubewells in U.P. are no exception in this regard. The financial plight of the State tubewells is summed up in Table 11.4. It is estimated that two-thirds of the revenue gap would remain unabridged even if all the tubewells had worked to their capacity output of 3,800 hours per year.[6] This gap could be covered only by raising tubewell tariff by 66 per cent. If that were done, the irrigators were not likely to cut-back their demand for tube-well water, as the demand is not found to be elastic with respect to the price of tubewell water. In point of fact, our hypothesis is that the utilization of irrigation potential of the tubewells would have been better than what has actually prevailed. For, a concern that does not break-even financially is forced to neglect routine maintenance of plant and machinery, giving rise to untimely breakdowns. Because of paucity of funds, prompt repair is not forthcoming. Therefore, a correct price policy can remove bottlenecks on the supply side. The farmers, especially their political leaders, must realize that under-pricing of water ultimately militates against their long-term as well as short-term interests. While financial losses eat into investible funds, untimely

5. It comprises of Agra, Aligarh, Badaun, Etah, Etawah, Farrukhabad, Kanpur, Mainpuri and Mathura. *Source*: Planning Commission, Resource Development Regions and Divisons of India, Delhi, 1964.
6. The basis of arriving at this capacity running hours is fully discussed in a larger study by this author: Under-utilization of Minor Irrigation Works: A Case Study of State-owned Tubewells in U.P. prepared at the Institute of Economic Growth, Delhi, 1969 (mimeo.).

and prolonged stoppages can be ill-afforded in agriculture where timeliness is the essence of a service.

TABLE 11.4

Financial Results of State Tubewells : 1946-47 to 1966-67

(Rs. crores)

Year	*Capital outlay*	*Gross revenue*	*Working expenses*	*Net revenue (+) or deficit (-)*	*Simple interest during the year*	*Net gain (+) or loss (-)*
1946-47	2.664	0.521	0.588	-0.067	0.084	-0.151
1947-48	2.912	0.696	0.673	+0.023	0.095	-0.072
1948-49	3.214	0.533	0.742	-0.209	0.104	-0.313
1949-50	3.932	0.891	0.828	+0.063	N.A.	N.A.
1950-51	4.491	0.987	0.914	+0.073	0.141	-0.068
1951-52	4.633	1.185	1.179	+0.006	0.150	-0.144
1952-53	7.124	1.012	1.145	-0.133	0.208	--0.341
1953-54	9.684	1.085	1.137	-0.052	0.312	--0.364
1954-55	13.055	1.254	1.295	-0.041	0.340	--0.381
1955-56	20.975	1.136	1.564	-0.428	0.587	-1.015
1956-57	24.766	1.285	1.989	-0.704	0.787	--1.491
1957-58	27.704	1.335	2.409	--1.074	0.930	--2.004
1958-59	28.218	1.706	2.862	--1.156	1.012	--2.168
1959-60	29.686	1.814	3.404	--1.590	1.059	--2.649
1960-61	31.180	2.103	3.809	--1.706	1.125	--2.831
1961-62	33.160	2.475	4.318	--1.843	1.211	--3.054
1962-63	36.694	2.134	5.329	--3.195	1.341	--4.536
1963-64	41.333	3.431	5.679	--2.248	1.521	--3.769
1964-65	46.171	4.297	6.430	--2.133	1.730	--3.863
1965-66	53.136	4.430	7.825	--3.395	1.992	--5.387
1966-67	58.656	4.647	8.610	--3.963	2.270	--6.233

Source : The Irrigation Department, Uttar Pradesh.

(3) The negative association between normal rainfall and demand for tubewell water suggests that extension of irrigation facilities in areas of high rainfall is bound to lead to somewhat lower utilization of irrigation potential. Thus, irrigation rates in relatively high rainfall

areas have to be relatively high. The only justification for having uniform rates could be either on constitutional grounds or economic backwardness of these high rainfall areas.

(4) The elastic relation between deviation from normal rainfall and demand for irrigation water suggests that large annual fluctuations in revenue receipts of a project can be dampened only by introducing an element of fixity in water rates. The introduction of a two-part tariff for tubewell water is a step in the right direction.[7]

7. The case for two-part tariff is strengthened by the cost structure of the irrigation industry which is dominated by fixed costs.

12

Unutilized Potential of State Tubewells in Uttar Pradesh*

Tubewells are destined to play a vital role in the development of agriculture in India. The recent experience has been that they are the most reliable source of irrigation during drought. For, all other sources, major or minor, are highly vulnerable to deficiency in rainfall. Furthermore, the future irrigation needs have perforce to be met increasingly from deep underground aquifers as the economical surface water resources have already been tapped substantially. The State has to play a major role in this exploitation, first in locating and mapping out underground water resources, and then making necessary investment in installation of tubewells, which is beyond the reach of majority of our farmers.[1]

The central objective in this short paper is to assess the magnitude and nature of the underutilization of capacity of State tubewells in U.P., a State which is a pioneer in developing this important source of minor irrigation in the country.[2] A part of the contribution of the paper, however, lies in the methodology innovated for measuring the unutilized potential -- a subject of great conceptual fascination and practical importance but operationally difficult to tackle. The need for departing from the conventional approach to the problem is briefly discussed in Section I. Then the new approach and the results flowing from it are spelt out in the following Section II of the paper.

---

* This paper is a part of the larger study Underutilisation of Minor Irrigation Works : A Case Study of State-owned Tubewells (IEG mimeo, 1969). It was published in Artha *Vijnana*, Vol. 15, No. 3, September 1973.

1. The estimated cost of a State tubewell in U.P. is about Rs. 1 lakh.
2. The development dates back to the mid thirties.

The basic data for the paper were furnished by the Office of the Chief Engineer, Irrigation Department, U.P., for which the author is grateful to the said Department.[3]

I

It is argued here that the basic data in respect of irrigation potential as well as area irrigated are unsatisfactory for assessing the unutilized potential of the State tubewells. If one were to employ the acreage data complied by the Irrigation Department, a picture of almost full utilization of capacity can emerge. This rosy picture, however, gets tarnished when one juxtaposes the State Revenue Department's statistics of area irrigated with the Irrigation Department's assessment of irrigation potential of the tubewells. These facts poignantly emphasize the hazards of using acreage data of the Irrigation Department in any assessment of the unutilized potential of an irrigation project in general and the State tubewells in particular.

Two sets of statistics about area irrigated by the State tubewells are listed in Table 12.1. Invariably, the statistics compiled by the Irrigation Department are far above these compiled by the Revenue Department. The main cause of this baffling phenomenon, according to a committee of experts, is the difference in recording procedures adopted by the two Departments.[4] Under the Irrigation Department's procedure, area of crops which receives tubewell water for more than one season is counted more than once. Accordingly, the expert committee endeavoured to demonstrate how the disagreement narrowed to barely one per cent by just making adjustment for the double-counting of sugarcane, the most important annual crop that spans over both the Kharif and the Rabi seasons. The author, however, is not wholly convinced by the able demonstration by the committee. For, the demonstration was for the year 1961-62, a year when the difference between the two sources was the lowest ever recorded, namely, 31 per cent. For subsequent years the disagreement still remains significantly wide even after adjusting for the double-counting of area under sugar-

3 The data were collected for my larger study entitled *Underutilization of Minor Irrigation Works: A Case Study of State-owned Tubewells in U.P.*, prepared at the Institute of Economic Growth, Delhi in 1969 (mimeographed).

4. The title of this committee's report is "Report of the Committee for Improving the Statistics of Irrigation in U.P." (unpublished.).

cane. For instance, the double-counting of sugarcane acreage reduces the 1965-66 difference from 63 to 41 per cent only.

TABLE 12.1
Extent of Disagreement in Statistics of Area Irrigated by State Tubewells

('000 Acres)

Year	*Area irrigated according to* *Irrigation Department*	*Revenue Department*	*Percentage difference of Col. 2 over Col.3*
1	*2*	*3*	*4*
1953-54	903	645	40
1954-55	1,098	717	53
1955-56	1.129	759	49
1956-57	1,157	820	41
1957-58	1,694	986	72
1958-59	1,590	1,023	55
1959-60	1,898	1,156	64
1960-61	1,811	1,223	48
1961-62	1,595	1,221	31
1962-63	2,004	1,327	51
1963-64	2,317	1,411	64
1964-65	2,394	1,537	56
1965-66	2,925	1,782	64

Sources : (1) For column 2, Irrigation Department, U.P.
(2) For column 3, *Basic Statistics Relating to Uttar Pradesh Economy* : 1950-51 to 1966-67, Economic and Statistics Directorate, Uttar Pradesh, June 1968 (mimeographed).

Let us now look into some estimates of irrigation potential. The State Irrigation Department publishes two sets of estimate of irrigation potential: (1) area irrigable on completion of the project and (2) irrigable area at present (see Table 12.2). Obviously, estimate (1) is

TABLE 12.2
Irrigation Potential of the State Tubewells

Year	*No. of tubewells in operation*	*Area irrigable by complete project (thousand acres)*	*Area irrigable at present (thousand acres)*
1947-48	2,062	1,122	1,046
1948-49	2,131	1,136	1,029
1951-52	2,393	1,232	1,232
1952-53	2,554	1,232	1,232
1953-54	2,931	1,579	1,579
1954-55	3,586	1,931	1,931
1955-56	4,515	2,640	2,035
1956-57	5,125	2,757	2,241
1957-58	5,806	2,931	2,241
1959-60	6,439	4,955	2,039
1960-61	6,603	2,892	2,039
1961-62	6,794	3,315	2,269
1962-63	7,129	5,404	2,334
1963-64	7,698	5,404	2,334

Source : Annual issues of Administration Report of the Irrigation Department, U.P.

not relevant for our purpose of enquiry. But the more relevant estimate (2) also raises some question-marks. First, why should the estimate be revised downwards in 1959-60 when the number of tubewells in operation was rising ? Second, why is its rise during 1947-63 not commensurate with several fold increase in the number of tubewells in the same period ? To put the same question differently, why does the current irrigation potential per tubewell decline from 507 acres in 1947-48 to 303 acres in 1963-64 ? Knowing that all State tubewells are of the same rated capacity, viz. 30,000 gallons per hour, what is the meaning and significance of a decline in potential of the order of 35 per cent over a 16-year period ? Has the calculus underlying the basis of estimating irrigation potential undergone a change over time ? Or, is the decline indicative of erosion of capacity due to poor maintenance ?

Irrigation engineers/administrators may justifiably revise their estimates of potential downwards if they find that the irrigators are prone to irrigating their lands more intensively than that assumed by them in their calculus. In this connection, it must be stated that the phenomenon of over-irrigation, so commonly reported in the case of canals, must be ruled out on *a priori* reasoning. For, tubewell tariff is tied to volume of water, and not to the crop.

II

If one were to tackle independently the task of measuring the level of capacity utilisation of the tubewells, one possible but practical way open is to take recourse to the Wharton School Method which has been employed in certain industrial studies. The essential effort in this method is directed at measuring the maximum sustainable level of output an industry can attain within a short time provided demand for its product is not a limiting factor.[5] Broadly speaking, the capacity output is identified through what is generally known as "trend-through-peaks" technique, that is, a curve is made to pass through the peak points of past record of output data. Obviously, this method can be used where past record of output of an irrigation work exists. The State tubewells in U.P. do satisfy this condition admirably, for their development dates back to the mid thirties.

It is evident that where the gap between the physical and the economic capacity outputs is likely to be large, the Wharton Index, with its practical aspect, has great relevance. While applying this technique to a public irrigation work, the unit of measurement of output of an irrigation work must be chosen appropriately. Evidently, volume of water is superior to irrigated area, for this unit is invariant with respect to intensity of irrigation, crop pattern, or the ultimate use of water of an irrigation work. Annual water output of the State tubewells is a simple product of their aggregate annual running hours and their water discharge rate. Since all the tubewells in U.P. are of the same discharge rate, their aggregate annual running hours can play the role of a perfect proxy variable for water output of the tubewells.

The highest peak running hours ever recorded in the thirty-year old history of the State tubewells was in the financial year 1944-45, when the average annual running performance of the 1,649 tubewells in

5. Klein, Lawrence R. and Robert Summers, *The Wharton Index of Capacity Utilisation*, Wharton School of Finance, University of Pennsylvania, Philadelphia, 1966.

operation at the beginning of this year worked out to be a little more than 3,900 hours. With a view to smoothening out any cyclical element in the performance, a five-yearly moving average of the data is taken, which reveals that the highest peak performance of the tubewells was attained in 1951-52, when the average for the 2,300 tubewells in operation in the year came to about 3,650 hours. Before taking this peak performance as indicative of capacity output for subsequent measurement, it may be desirable to make an upward adjustment on account of the fact that a few tubewells were in formative, or development stage.[6] Since about one-tenth of the tubewells in operation in 1951-52 were in this development stage, it is proposed to take 3,800 hours as the capacity output of a State tubewell.[7]

To make sure that 3,800 hours capacity output is not much off the mark under the economic framework of the sixties, an attempt is made to study the peak performance of the tubewells at a disaggregative level. The highest average running hours per tubewell per year during the five-year period 1960-61 to 1964-65 was about 3,924 for Etawah district in Western U.P. However, this exceptional performance may be ignored on the simple consideration that there was just one State tubewell in operation in this district between 1960-61 and 1964-65. The next highest performance for any district was 3,438 hours for the western district of Muzaffarnagar. Though this performance is somewhat short of the capacity performance of 3,800 hours, the difference is accountable by the formative stage through which many a tubewell was passing during 1960-65 (to be precise, 19 per cent in 1960-61 and 20 per cent in 1964-65).

How does the 3,800 - hour capacity mark compare with that financial point for a tubewell where revenue and work costs just break

6. A tubewell is said to be in development stage so long as its life from the date of its energisation does not exceed three years. Presumably, a commissioned tubewell is considered to be in formative stage so long as either (a) the full complement of its water courses is not constructed or (b) the water courses are not fully lined, or both. Delays in (a) can occur because of difficulty in acquiring land from farmers, wheras delay in (b) may be on account of shortages in lining materials, lack of synchronisation in constructing different parts of a tubewell etc. From the viewpoint of suppply, a commissioned tubewell is not fully complete unless the network of its water courses that carry water from the pumping point to the farmers' field channels, is properly and fully executed.
7. In effect, we are assuming that the level of utilisation of the tubewells in development stage was half of that of the other tubewells.

even ? On the basis of the operating costs and tubewell tariff prevailing at the end of the fifties, the financial break-even point was placed at 3,200 hours by the Minor Irrigation Team of the Committee on Plan Projects.[8] In the wake of the withdrawal of the three-anna-in-a-rupee rebate in April, 1961, the effective price of tubewell water went up by 23 per cent. But the sixties are characterised by a sharp upward trend in general price level, notably since the beginning of 1964 when inflationary pressures emerged in the whole economy. It appears that 3,200 hours per year could be broadly taken as the financial break-even point for a tubewell upto the end of the Third Plan period, but not beyond. Probably, it is now very near the 3,800-hour mark.

For the year 1967-68, the running hours per tubewell averaged at 2,536. With capacity output taken at 3,800 hours, the extent of underutilisation per tubewell works out to be about 1,264 hours, that is, almost one-third of the productive capacity of the tubewells remained unutilised in 1967-68. Thus, a considerable part of the tubewell potential for irrigation was wasted in 1967-68, thereby rendering a part of the total capital investment of the order of Rs. 600 million in the State tubewells infructuous.[9]

Underutilisation of capacity can be traced either to demand factors or supply factors, or both. On the demand side, benefit-cost ratio for tubewell water, depending in turn upon a number of variables -- such as the price of tubewell water, the level of agricultural prices and the marginal product of tubewell water -- is an important factor to be reckoned with. The unusually high level of agriculture prices and the introduction of the HYV programme would be strong factors in strengthening the demand for tubewell water in 1967-68. The only factor on the demand side that would have tended to weaken these strong demand forces would be the occurrence of good and timely rains which indeed

8. 4,500 hours if interest charge is deemed as part of the total cost. See, *Report on State Tubewells*, New Delhi, 1961, p. 17.

9. This estimated capital investment is exlusive of the huge cumulative arrears of simple interst (about Rs. 180 million) which could not be provided for (and are normally included in the term capital-at-charge) in any year since the beginning of 1952-53. What is equally important to bear in mind is that this capital investment is at book value only. On the basis of bench-mark information about capital costs for construcing tubewells it is estmated by the author that the replacement cost of the tubewells was about Rs. 1170 million. It is this figure of capital investment that gives a real idea about the extent of underutilisation of capital resources.

was the case in this year. Therefore, a small part of the underutilisation has to be attributed to this transitory factor, namely, good weather.

On the supply side, the capacity of the tubewells to meet the required demand for water would be reduced by three factors, namely, (a) the proportion of the tubewells in development stage, (b) the depreciation of their rated capacity and (c) the inability of the Irrigation Department to provide timely repairs in the event of break-downs. Taking up these factors, one by one, about 11 per cent of the total of 8,699 tubewells in operation in 1967-68 were in the development stage. Thus a part of the underutilisation has to be ascribed to this transitory factor.

The effective capacity of tubewell is primarily governed by (1) the water discharge rate, (2) the state of the water courses or tubewell "guls" and (3) the ability of the tubewell machinery to function continously during the agricultural seasons, notably during peak part of each season. When the discharge rate comes to fall below its rated capacity, the effective irrigable potential falls. Likewise, if the "guls" are in a bad shape from the point of view of repair and maintenance, this vital link in the chain can lead to a reduction in effective potential, even if the discharge rate is unaffected over time. Finally, frequent mechanical breakdowns of the tubewell machinery during agricultural season are clear pointer to the reduction in the rated capacity of a tubewell.

There is no doubt that time takes its toll of the capacity of any capital asset. This happens in spite of repair and maintenance work. However, the quality of repair/maintenance work mainly determines the rate of decline of the productive efficiency, as well as the point on time path whence the declining phase of capacity begins. Routine, regular maintenance, that is so vital in the efficient and smooth functioning of all mechanical assets, not only defers the above point but also reduces and regularises the rate of decline during the declining phase of capacity. The same applies to timely and adequate repair work. Thus, both natural depreciation due to age factor and premature depreciation of capacity due to either insufficient routine maintenance or inadequate and untimely repair, are responsible for diminished effective capacity of a tubewell.

How far the effective capacity of the State tubewells in 1967-68 was below the rated one can be possibly unearthed if we hold the demand factors constant while analysing a frequency distribution of the running performance of these tubewells. It so happens that such a frequency distribution is readily available for the immediately preceding year 1966-67, a year of intense drought. Since 1965-66 was also a year of widespread drought, it is presumed there was no deficiency of demand for tubewell water in 1966-67. Thus, low performance of any tubewell in this year must be attributed to bottlenecks on the supply side only.

Out of the 8,271 state tubewells in operation in 1966-67, as many as 7,041 were reported to be positively out of their development phase in this year, that is, their age was four years and above. The frequency

TABLE 12.3
Frequency Distribution of Tubewells according to Running Hours During 1966-67

Running Hours	*No. of Tubewells*	*Percentage of Tubewells*	*Cumulative Percentage*
0 to 100	29	0.41	0.41
101 to 200	16	0.23	0.64
201 to 300	10	0.14	0.78
301 to 400	17	0.24	1.02
401 to 500	13	0.18	1.20
501 to 1000	118	1.68	2.88
1001 to 1500	285	4.05	6.93
1501 to 2000	446	6.33	13.26
2001 to 2500	685	9.73	22.99
2501 to 3000	807	11.46	34.45
3001 to 3500	1005	14.27	48.72
3501 to 4000	1066	15.14	63.86
4001 to 5000	1854	26.33	90.19
5001 to 6000	633	8.99	99.18
Above 6000	57	0.82	100.00
Total	7041	100.00	100.00

distribution of such tubewells is set out in Table 12.3. Of the seven thousand tubewells under reference, a little more than 50 per cent operated for over 3,500 hours during 1966-67. Out of these 50 per cent, about one half showed a running performance of 4 to 5 thousand hours, while another sizable fraction (one-sixth) came in the high range of 5 to 6 thousand hours, and a little fraction even overshooting the 6 thousand hours mark--just 30 per cent short of the 8,760 hours that are there in a single year. These excellent performances are perfectly understandable when we bear in mind two points : (a) that the farmers' own sources of irrigation within the rather large culturable command[10] of a tubewell must have been awfully depleted due to two successive droughts, and (b) the use of tubewell water for drinking purposes[11] was quite significant as many of the sources of drinking water also dried up. In spite of such favourable demand conditions, not to mention of the tremendous improvement in benefit-cost ratio of irrigation water, a large number of tubewells showed rather unsatisfactory performances.

The supply side has two aspects viz., external and internal. Among the external factors are the supply of power and the state of law and administration. If supply of power is reduced, the performance of a tubewell is reduced. Likewise, thefts of vital tubewell installations, such as copper coil of transformer and electric motor, seriously impair the working of the tubewells, for these may not be so easy to repair or replace promptly. While it has to be granted that these external factors have some bearing on the performance of the tubewells in 1966-67, the unusually low performance of many tubewells in 1966-67 cannot be wholly attributed to them. It is argued here that the low performance was partly due to the inability of the tubewells to operate continuously under peak demand conditions.

It is but natural that repair and maintenance work becomes untimely and inadequate when a concern is operating under financial losses year after year. In fact, even routine maintenance work, that can be undertaken at a nominal cost so as to avert major repair work in future, gets neglected in a losing concern that does not recover even its running costs. (At the top of this, the cost of repair and maintenance as well as the replacement cost has been steeply mounting up for all

---

10. The cultural command of a State tubewell is indeed quite big in relation to the net area irrigated by a tubewell, because even the ratio of command area to gross irrigated area is about three.
11. Many tubewells are, in fact, equipped with the necessary appurtenant works, such as, overhead water storage tanks and outlet channels, for rural water supply.

capital assets in general and the tubewells in particular). Thus, part of the decline in the rated capacity of the tubewells may be traceable to the bad financial plight of the State tubewells.

CONCLUSIONS

The upshot of the preceding probe in the nature of the underutilisation of the capacity of the State tubewells is that only a part of the unutilised capacity can be described as a transitory or a passing phenomenon. To a large extent, this type of underutilisation is inevitable in an activity, which is expanding with time, whose product is demanded in definite weather cycles, and which is suffering due to bad law and order situation, particularly in the countryside.[12] However, a significant part of the underutilisation is due to significant erosion in the rated capacity of the tubewells. In the ultimate analysis, the erosion in capacity can be traced to the serious under-pricing of tubewell water.

12. Incidence of thefts of transformers and electric motors is reported to be comparatively high in the western districts which are in proximity to the brassware industry of Moradabad.

13

Census of Minor Irrigation : Some Results

The definition of minor irrigation work/scheme today is one whose 'culturable command area'(CCA) is upto 2,000 hectares. This area-based or physical definition was adopted by the Indian Planning Commission in April 1978 because it need not be revised periodically in the wake of persistent price inflation in the economy--earlier, the financial definition, adopted by the Indian planners at the time of the First Five Year Plan (1951-56), had to be modified in March 1965 (unit investment upto Rs. one million raised to Rs. 1.5 million) and, again, in March 1970 (upto Rs. 2.5 million for the plains and upto Rs. 3.0 million in hilly regions). It may be noted that in the pre-plan era the definition of a minor irrigation work was indeed in area terms, though irrigable potential was measured in irrigated area, not in the CCA of a minor irrigation work. Anyway, all groundwater-based irrigation structures, such as dugwells and tubewells, are now deemed to be in minor irrigation segment.

TYPES OF MINOR IRRIGATION WORKS

Since April, 1978 all ground water schemes and surface water schemes (both flow and lift) having culturable command area upto 2000 hectares (ha.) individually are considered as minor irrigation works. Ground water schemes comprise dugwells, deepening of dugwells, dug-cum-borewells, borings, private shallow tubewells, filter points, deep tubewells etc. The command area of private ground water scheme varies from 1 to 5 ha. The State Governments provide assistance in installation of such schemes which confines mainly to technical guidance, custom service for boring and arrangements for credit facilities at reasonable rates of interest. The subsidies are also made available for installation of these schemes to weaker sections of farmers. The construction, operation and maintenance of these schemes are done by the farmers themselves. The deep tubewells which extend upto the depth of 200 metres or more are designed to give a discharge of 100 to 200 cubic metres per hour. These schemes are usually beyond the means of individual farmers and their construction, operation and maintenance are done by the State Government Departments or Corporations.

The surface water schemes comprise surface flow schemes and surface lift irrigation schemes. The surface water flow schemes are: tanks, check dams, and diversion channels which are sometimes temporarily built. The diversion channels are generally prevalent in hilly regions. The names of these schemes vary from State to State. These schemes are known as *Kuhls* in Jammu & Kashmir and Himachal Pradesh, *Guls* in Uttar Pradesh and *Aharas and Pynes* in Bihar and low *Khones and Dongs* in North Eastern region. While the small storage tanks are owned by community or local bodies and generally have command area upto 40 ha, the large storage tanks, alongwith the distribution system, having command area varying from 40 to 2000 ha. are constructed by State Government Departments.

Minor irrigation has come to be associated with admirable properties like low-gestation period and low development cost, whereby a sort of halo has come to pervade around this mode of irrigation[1]. In contrast, critical scrutiny of claims made on behalf of major irrigation schemes (CCA above 10,000 hectares)—especially in which big dams play a pivotal role—has greatly tarnished the image of large scale irrigation works. No wonder, then, one comes across pleas for exclusive development of minor irrigation sources in Indian irrigation planning *e.g.,* B.B. Vohra pleads for a freeze on major irrigation during the Ninth Five Year Plan (1997-2002).

It is high time that minor irrigation is subjected to a critical appreciation free from emotional tinges. It is proposed to assess the ground reality in this regard mainly with the help of *Report on Census of Minor Irrigation Schemes 1986-87* released in 1994 by the Ministry of Water Resources (MOW)[2].

---

1. Ten dimensions of such a halo are enunciated in the *Report of the Working Group on Minor Irrigation for Formulation of the Ninth Plan Proposals* (see pp. 15-16).
2. It is a printed, hard-bound document but not a priced one. It is brought out by the Minor Irrigation Division of the MOW.

NEED FOR A CENSUS

Minor irrigation development in the States is dealt by different departments under different developmental sectors. The progress reports relating to minor irrigation programmes received form different departments at the Centre are used for compilation of minor irrigation data for the State. There has not been a single nodal department in the State to compile the minor irrigation statistics for the entire State. The structures installed by the farmers from their own efforts are generally not firmly recorded. The irrigation potential of ground water schemes are usually reported on the basis of certain assumed yardsticks. Usually no efforts have been made at the State level to provide a scientific basis for such yardsticks. In the case of State irrigation works the irrigation potential and utilisation are usually estimated on the basis of assumed water duties/water allowances. The reported figures of irrigation potential created by certain schemes during a particular period might have depreciated over a period of time but no depreciation is taken into account in reporting the figures of cumulative irrigation potential created. The figures in respect of private minor irrigation works installed totally by the efforts of the farmers are usually not based on any positive enumeration or sample survey.

In certain States there are no field agencies in the minor irrigation department to collect data relating to minor irrigation programme for administrative reports. Further, since minor irrigation programme is dealt by more than one department of the State, the administrative report of irrigation department gives the statistics pertaining to works maintained by the department.

During the discussions of annual plan proposals of the States hardly any State gives a clear picture of the development of minor irrigation in the State as a whole. A sub-committee on irrigation statistics set up by the Planning Commission in 1970, therefore, recommended a detailed census of minor irrigation works to strengthen the data base with regard to minor irrigation development in India. The National Commission on Agriculture endorsed this, adding that a census of irrigation sources be undertaken alongwith the Agricultural Census once in five years. Special irrigation surveys on the number of works and their utilisation might be undertaken by the States. The Technical Committee on Agricultural Census

1980-81 discussed the list of items in respect of data relating to minor irrigation proposed to be collected alongwith the agricultural census in 1980-81. It was felt that the agricultural census data are being compiled mainly from the existing land records. Since the required information on minor irrigation does not figure in the land records, it would not be possible to collect this information through the main agricultural census.

The information on area irrigated is an ancillary product from the Land Use Statistics (LUS). The Directorate of Economics and Statistics, Department of Agriculture, Government of India, publishes regularly the national level information with a time lag of about 3 to 4 years. The primary data for LUS are collected by village *patwaries* through plot-by-plot enumeration in many States, but are estimated on the basis of sample surveys in other States (notably West Bengal, Orissa and Kerala). The basic enumeration forms are not the same in States and the instructions for obtaining details of area under irrigation and its sources are inadequate. Classification of sources of irrigation does not provide data on irrigated area by important types like private tubewells, dugwells, deep tubewells, diversion schemes and surface lift irrigation schemes. The gross irrigated areas according to different sources in most of the States are also not being compiled in LUS. In case crop area is irrigated by two different sources, say, canal and well, it is usually recorded under canal head. Also, the LUS data do not give the number of minor irrigation structures.

The Scheme for Improvement of Irrigation Statistics was launched in 1980-81 by the Department of Agriculture. Although a census of minor irrigation works was to be taken up under this scheme, it could not be conducted till 1985-86. While discussing various issues and problems in the irrigation sector in the Planning Commission in 1986, it was stressed that census of minor irrigation schemes be conducted within a period of one year. Thus commenced the first census of minor irrigation, with 1986-87 as the reference year. In Appendix Note the methodological details of this census are spelt out.

The census of minor irrigation data, though dated in time, adds to our information in two ways:(1) as a crosscheck on presently available estimates in this area and (2) as a source of information hitherto not known, or known vaguely, for the State or district as a whole. To illustrate, one example of type (1) is the number of minor irrigation works of various types that were in existence by the end of 1986-87, whereas the number of such works that have gone out of use--say, because of cumulative neglect of repairs and preventive maintenance (as in the case of tanks) and due to technological obsolescence (dugwells falling into disuse because of new tubewell technology)—would be an example of type (2). In what follows we present the outcome of some exercises which the author has conducted on the census data. We begin with a crosscheck on the Planning Commission's estimates of minor irrigation capacity.

I. ESTIMATES OF MINOR IRRIGATION CAPACITY

At the state level, it is not that easy for a state government to know quickly and accurately the progress in minor irrigation development in its own state. For one, much of the development is outside the public sector domain, notwithstanding that some state/official body may be involved in development of minor irrigation within private ownership. For another, besides the state department of minor irrigation, other departments implementing development programmes aimed at rural development, or poverty alleviation, or employment generation, or tribal welfare, do carry out works relating to minor irrigation.

Till the conduct of census of minor irrigation works, the only source of information about created minor irrigation capacity has been the Plan documents released periodically by the Planning Commission. The Plan estimate is based on progress reports on minor irrigation furnished by the state governments while seeking approval of the Commission for their respective annual as well as five year plans. The states, in turn, compile these statistics on the basis of progress reports received or information gathered from their district planning units which in turn depend on information supplied by different district level government departments and financial institutions associated with the construction of various categories of minor irrigation structures. In addition, annual reports of Rural Electrification Corporation (REC) and State Electricity Boards (SEBs) provide information on energisation of pumpsets so widely used in minor irrigation works these days.

The principal advantage of this reporting system is one of timely compilation of the needed data for purposes of decision making at the stage of annual plan making. So, while planning in the year 'T' for the subsequent year 'T+1', the central planning body would have before itself the following data for a state seeking plan approvals.

a. *actual* achievements in establishing new minor irrigation works (by type and by irrigable area capacity) for the preceding year T-2;
b. *likely* achievements in establishing new minor irrigation works (by type and by irrigable area capacity) in the preceding year T-1;
c. *anticipated* achievements of the targets, by type and irrigable area capacity, in the ongoing year T;
d. targets proposed by a state for new capacity in minor irrigation for the year T+1.

At the time of plan approvals, it is practically impossible to check on the veracity of the estimates (a) and (b). For one, the village records, wherever maintained by a government functionary, are not required to be kept about irrigable area or irrigation capacity figures for various irrigation structures. For another, even information in respect of the number of minor irrigation structures like dugwells and tubewells established in farmers' own fields becomes available after considerable lag of time. This lag is usually above two years. So, the state planning bodies are better placed than the central planning body to overproject their side of the achievements on minor irrigation front. The motivation for such overprojection may not have much to do with getting more plan funds for minor irrigation development, because bulk of such development is taking placing under private aegis (*i.e.,* without any financial allocations in the state public sector plan). Though it is difficult to identify motives behind overprojection, yet two such motives may be stated here. One, much of the private investment in minor irrigation is financed by financial institutions under public control. Statewise allocations of limited institutional credit for financing agriculture do bear some relationship with programmes for development of minor irrigation works. Two, it may be simply a desire on the part of some overeager state politicians/officials to place the state performance in a brighter light.

The Census of minor irrigation provides a bench-mark data on very many aspects of minor irrigation in the country. This data pertains mostly to the year 1986-87 (for 1987-88 in the case of West Bengal). So, we can compare for this year the Plan statistics on irrigation capacity, as also on number of minor irrigation structures, with the corresponding Census statistics. We can thus ascertain the magnitude and the direction of difference between the two data sources. This is what has been attempted in what follows. Before we do that, it may be pointed out that a very feeble attempt in this direction was made by the Working Group on Minor Irrigation for the Ninth Plan (September, 1996). Without making any comment on the outcome of such an effort, its report simply records that the Census-based figure of established irrigation capacity (duly adjusted for the omitted state of Rajasthan) for minor irrigation segment as a whole was almost the same as that reported earlier in the plan document—some marginal differences in the case of groundwater as well as surface water components emerged, as shown below:

	Census estimate	*Plan estimate*
1. Groundwater potential	29.83 mha	29.08 mha
2. Surface water potential	8.71 mha	9.78 mha
3. Total minor irrigation potential	38.54 mha	38.54 mha.

Whether this reassuring picture was valid at the state level was not examined by the Working Group.

CAPACITY DIFFERENCES

At the aggregate level (excluding Rajasthan), the two estimates of total minor irrigation potential created by 1986-87 are quite comparable. Whereas the Plan sources reveal an estimate of the order of 39 million hectares (mha), the Census source places the same at about 42 mha. Thus the Plan estimate is lower than the Census estimate by about 8 per cent. This is indeed an unexpected finding, because the ground reality—as thrown up by the Census—was feared to be one of lower achievement than indicated in plan documents. Before we pursue this aspect, let us delve deeper via a disaggregative analysis.

When one distinguishes the potential between groundwater structures and surface water sources, one discerns a contrast in the direction of the aforesaid difference. In the case of groundwater potential, which is much larger than surface water-based potential, the Plan estimate is lower than the Census estimate by nearly 12 per cent. It is in the case of surface water that the Plan estimate is somewhat higher (6 per cent) than the Census figure. The two sets of statistics are listed below for ready comparison.[3]

	Plan estimate (mha)	*Census estimate (mha)*	*Plan estimate relative to Census estimate (%)*
Total* minor irrigation potential in 1986-87 from:			
Groundwater	28.72	32.49	88
Surface water	9.89	9.33	106
Total	38.61	41.82	92

* Excludes Rajasthan.

As we disaggregate further by states, we come across several cases of substantial divergence in the two sets of estimates, and fewer cases of concurrence. In Table 1 are set out these comparative statistics of total minor irrigation capacity by state. There are thirteen states where the difference is negative *a la* the national level, and there are twelve states for which the direction of the difference is the other way round, that is, the Plan figure exceeds the Census figure. One striking negative difference is for Gujarat state (minus 55 per cent), whereas one of the most substantial differences in the other direction is in respect of Uttar Pradesh (plus 38 per cent). These facts show that there is no uniformity either in the (relative) magnitude or in its direction when it comes to state level differential in the estimates of total minor irrigation potential emerging from the Plan and the Census sources. It is mere fortuitous circumstance that the plus and the minus state-level differences in capacity largely counterbalance in the national aggregate.

Let us see if the state level difference, if any, is uniform in direction and relative magnitude as between groundwater and

3. Discerning readers would notice discrepancies between this data set and the similar one cited earlier from the Report of the Minor Irrigation Group for Ninth Plan. However, factors behind this could not be ascertained.

Table 1
Comparison of Plan and Census Estimates of Total Minor Irrigation Capacity, 1986-87

S.No.	*State*	*Plan estimate (000ha)*	*Census estimate (000ha)*	*Difference:Col.3-Col.4 absolute (000ha)*	*relative* (%)*
(1)	*(2)*	*(3)*	*(4)*	*(5)*	*(6)*
1.	Andhra Pradesh	2,471	3,385	(-)914	(-)27
2.	Arunachal Pradesh	48	61	(-)13	(-)21
3.	Assam	452	241	211	88
4.	Bihar	3,787	3,142	645	21
5.	Goa	15	18	(-)3	(-)17
6.	Gujarat	1,728	3,844	(-)2,116	(-)55
7.	Haryana	1,423	1,938	(-)515	(-)27
8.	Himachal Pradesh	120	196	(-)76	(-)39
9.	Jammu & Kashmir	344	300	44	15
10.	Karnataka	1,195	1,740	(-) 545	(-)31
11.	Kerala	429	396	33	8
12.	Madhya Pradesh	2,170	2,388	(-)218	(-)9
13.	Maharashtra	2,076	2,648	(-)572	(-)22
14.	Manipur	43	21	22	104
15.	Meghalaya	38	50	(-)12	(-)24
16.	Mizoram	8	7	1	14
17.	Nagaland	55	52	3	6
18.	Orissa	1,170	827	343	41
19.	Punjab	3,219	4,772	(-)1553	(-)33
20.	Rajasthan	2,059	NA		
21.	Sikkim	17	16	1	6
22.	Tamil Nadu	2,008	2,988	(-)980	(-)33
23.	Tripura	66	60	6	10
24.	U.P.	13,847	10,003	3,814	38
25.	West Bengal	1,807	2,359	(-)552	(-)23
26.	Total (1 to 25)	40,595	NA	NA	NA
27.	UTs	76	101	(-)25	(-)25
28.	All-India	40,670	NA	NA	NA
29.	All-India excluding Rajasthan	38,611	41,824	(-)3,213	(-)8

* to Census estimate.

Table 2
Statewise Comparison of Minor Irrigation Capacity Estimates from Groundwater and from Surface Water

S.No.	State	GroundWater				Surface Water			
		Plan esti-mate (000ha)	Census esti-mate (000ha)	Difference : Col.3-Col.4 absolute (000ha)	relative* (%)	Plan esti-mate (000ha)	Census esti-mate (000ha)	Difference : Col.7-Col.8 absolute (000ha)	relative* (%)
1	2	3	4	5	6	7	8	9	10
1.	Andhra Pradesh	1.303	1.831	(-)528	(-)29	1,168	1,554	(-)385	(-)25
2.	Arunachal Pradesh	-	-	-	-	48	61	(-)12	(-)20
3.	Assam	110	52	58	112	342	189	-	-
4.	Bihar	2.531	2.394	137	6	1,256	748	-	-
5.	Goa	0.7	3.5	(-) 2.8	(-)80	14	14	(-)1	(-)7
6.	Gujarat	1.546	3.649	(-)2,103	(-)58	182	215	(-)33	(-)15
7.	Haryana	1.384	1.932	(-)547	(-)28	39	6	33	550
8.	Himachal Pradesh	12	19	(-)7	(-)37	108	176	(-)49	(-)31
9.	Jammu & Kashmir	6	6	0	0	338	294	17	5
10.	Karnataka	509	735	(-)226	(-)31	686	1,004	(-)318	(-)32
11.	Kerala	69	55	14	25	359	342	17	5
12.	Madhya Pradesh	1,266	1,662	(-)396	(-)24	903	726	177	24
13.	Maharashtra	1,311	2,285	(-)947	(-)43	765	904	(-)139	(-)15
14.	Manipur	-	-	-	-	43	21	-	-
15.	Meghalaya	9	11	(-)2	(-)18	29	39	-	-

(Contd...)

16.	Mizoram	-	0	-	-	8	7	2	33
17.	Nagaland	0.4	0	0.4	NC	55	52	3	6
18.	Orissa	544	234	310	132	627	593	34	6
19.	Punjab	3,184	4,765	(-)1,581	(-)33	35	7	28	400
20.	Rajasthan	1,678	NA	NA	NA	381	NA	NA	NA
21.	Sikkim	-	0	-	-	17	16	1	6
22.	Tamil Nadu	1,168	1,824	(-)654	(-)36	840	1,164	(-)164	(-)16
23.	Tripura	13	9	4	44	53	51	2	4
24.	U.P.	12,942	9,739	3,203	33	905	295	610	207
25.	West Bengal	755	1,236	(-)480	(-)39	1,053	1,123	428	68
26.	Total (1 to 25)	30,341	NA	-	-	10,254	NA	-	-
27.	UTs	60	77	(-)17	(-)22	16	(-)23	(-)7	(-)30
28.	All-India	30,401	NA	NA	NA	10,269	NA	NA	NA
29.	All India excluding Rajasthan	28,723	32,493	(-)3,770	(-)11	9,888	9,331	557	6

* to Census estimate.

surface water based irrigation capacities. This is attempted in Table 2. In the case of Gujarat state, the observed phenomenon of substantial excess of Census over Plan figure of total minor irrigation potential turns out to be mainly one pertaining to groundwater component, the dominant of the two components of state minor irrigation. More specifically, the Census figure for groundwater potential for this state exceeds the corresponding Plan figure by nearly 2 million hectares (mha). But in the case of Uttar Pradesh, the Census figure for groundwater turns out to be nearly 3 mha less than the corresponding Plan figure. It is this kind of offsetting differences at state level that give rise to a reassuring picture of agreement in two data sets at the national level.

In relative terms, the Gujarat difference between Plan and Census figures works out to about minus 58 per cent for groundwater potential and about minus 15 per cent for surface water potential. The plan estimate for surface water minor irrigation works in U.P. exceeds the corresponding Census number by nearly 0.6 mha. In relative terms, however, divergence here in the case of groundwater potential estimates is lower, namely, 33 per cent as compared to 207 per cent for surface water minor irrigation potential.

DIFFERENCES IN NO. OF STRUCTURES

Minor irrigation potential (P) is a product of two things:

(1) number of minor irrigation works (N) and (2) irrigation potential per minor irrigation work (A). That is to say, P = N.A.

Any disagreement between the Census and the Plan figures in respect of minor irrigation potential could be traced either to differences in 'N' or in 'A' (or both) in the two data sets. Fortunately, differences in respect of groundwater-based irrigation works in the two data sets can be investigated to shed light on statewise differences in groundwater irrigation potential between the Census and the Plan figures for 1986-87. In what follows this exercise is carried out at the national level, as also for the two states of Gujarat and Uttar Pradesh.

For irrigation purpose there are basically two types of groundwater structures, namely, dugwells and tubewells. In the case of tubewells, a distinction is made in the Plan reports between private tubewells and public tubewells. On the other hand, the Census distinguishes tubewells by depth only—that too in terms of two categories only, namely, shallow and deep tubewells. Though most public tubewells are deep tubewells, it is not so in the reverse case. That is to say, all deep tubewells are not necessarily public tubewells. This needs bearing in mind especially in the case of Gujarat where private enterprise (farmers as well as water companies) has invested in deep tubewells on a notable scale, so as to get bigger rate of water discharge. In view of this, the observed differences in the number of tubewells of a given type need to be interpreted with caution, more so in the case of Gujarat state.

Another allied statistic of some relevance in the context of groundwater potential is the type of water lifting mechanism. In the Indian conditions, a well fitted with an electric pumpset *ceteris paribus* tends to have greater irrigation capability than one fitted with an oil-driven pumpset, which in turn has more irrigable area potential than a well fitted with a traditional lift such as 'mhote', persian wheel or piccota. We have to bear in mind that all electric motors are not exclusively used in groundwater irrigation. Some of them are used in surface water lift irrigation works, more so in the states of Kerala, Karnataka, Maharashtra and Madhya Pradesh. According to the Census, barely 4 per cent of the total electric motors in the country were being used for surface water lift irrigation during 1986-87.

At the national level, there is some divergence in total number of wells (dugwells + tubewells): whereas the Census number is around 11.7 millions, the Plan number is about 12.2 millions, about 0.5 million more than the Census figure. Though the number of dugwells in the Plan documents exceeds the one in the Census report by a margin of 6 per cent, the direction of the difference in respect of tubewells is other way—the Plan number turns out to be about 3 per cent lower than the one brought out by the Census report. It is the Plan number of electric motors (6.4 millions) that substantially exceeds the Census number of 4.9 millions, a difference of 1.5 million motors (31 per cent). The comparative picture of number of dugwells, tubewells and electric motors is set out below:

		Plan estimate	*Census estimate*	*Plan estimate as per cent Census of estimate*
1.	No. of dugwells (millions)	8.2	7.7	106
2.	No. of shallow tubewells (millions)	3.9	4.0	97
3.	No. of deep tubewells (000)	53	104 (81 if Gujarat is excluded)	51 (65)
4.	No. of electric motors (millions)	6.4	4.9	131

The Plan number exceeding the Census number by 31 per cent in the case of electric motors is overstated (by about 5 percentage points) owing to our having made an implicit assumption that all electric motors used in minor irrigation works are used by well owners only. As already mentioned, the Census shows about 96 per cent of such motors are used in well irrigation, and the rest 4 per cent in surface lift irrigation schemes. Obviously, the Plan figure about number of electric motors, based as this is on data furnished by the State Electricity Boards, is on the high side by a factor of one-fourth. In fact, Sant and Dixit (1996) too found that in Maharashtra the SEB statistics do reveal significant overprojection of ground position in respect of electric motors used in minor irrigation works.[4]

As one comes to state level, the high level of concurrence between the Census and the Plan numbers for wells tends to vanish (vide Tables 3 to 5). Let us explore with the help of data for U.P. and Gujarat--the two states exhibiting the marked disagreement in irrigation capacity figures--how for the disagreement is due to corresponding difference in number of wells. It is our contention that 33 per cent difference in minor irrigation capacity in the case of U.P. is mainly due to the Plan estimate of number of dugwells exceeding the Census estimate, and much less due to the difference in number of tubewells.

Tubewell is the dominant mode of well irrigation in U.P. According to the Census, the state's irrigation capacity from shallow

4. For more details, see 'Analysis of Electricity Consumption of Irrigation Pumpsets in the State of Maharashtra,' report prepared by Girish Sant and Shantanu Dixit for PRAYAS, an NGO from Pune, 1996.

Table 3
Comparative Number of Dugwells Constructed upto 1986-87

S.No.	States	*No. of dugwells(000)* Plan	Census	*Difference: col.3-col.4* absolute (000)	relative* (%)
1	*2*	*3*	*4*	*5*	*6*
1.	Andhra Pradesh	999	1,175	(-)176	(-)15
2.	Arunachal Pradesh	0.4	-	-	-
3.	Assam	-	1	1	-
4.	Bihar	366	338	28	8
5.	Goa	0.4	5	(-)4.6	(-)92
6.	Gujarat	684	839	(-)155	(-)18
7.	Haryana	42	39	3	8
8.	Himachal Pradesh	2.5	2	0.5	27
9.	Jammu & Kashmir	0.2	1	(-)0.8	(-)80
10.	Karnataka	495	406	89	22
11.	Kerala	158	124	34	27
12.	Madhya Pradesh	1,196	1,118	78	7
13.	Maharashtra	1,037	1,390	(-)353	(-)25
14.	Manipur	-	-	-	-
15.	Meghalaya	0.1	Neg.	Neg.	Neg.
16.	Mizoram	-	-	-	-
17.	Nagaland	0.8	Neg.	Neg.	Neg.
18.	Orissa	507	227	280	123
19.	Punjab	93	21	72	342
20.	Rajasthan	814	na	-	-
21.	Sikkim	-	0	-	-
22.	Tamil Nadu	1,433	1,490	(-)57	(-)4
23.	Tripura	-	1	675	46
24.	Uttar Pradesh	1,136	461	675	146
25.	West Bengal	33	64	(-)31	(-)48
26.	Total (1 to 25)	8,997	NA	NA	NA
27.	UTs	16	6	11	166
28.	All India	9,013	NA	NA	NA
29.	All-India excluding Rajasthan	8,199	7,710	489	6

* to Census figure; Neg.—negligible; NA—not available.

Table 4
Comparative Number of Shallow Tubewells Constructed upto 1986-87

S.No.	*States*	*No. of shallow tubewells (000)*		*Difference: col.3-col.4*	
		Plan	*Census*	*absolute (000)*	*relative* (%)*
1	*2*	*3*	*4*	*5*	*6*
1.	Andhra Pradesh	89	110	(-)21	(-)20
2.	Arunachal Pradesh	-	0	-	-
3.	Assam	27	16	11	69
4.	Bihar	501	424	77	18
5.	Goa	-	Neg.	Neg.	Neg.
6.	Gujarat	5	5	0	0
7.	Haryana	402	357	45	13
8.	Himachal Pradesh	1	1	0	0
9.	Jammu & Kashmir	2	1	1	100
10.	Karnataka	23	40	(-)17	(-)43
11.	Kerala	1	0	1	NC
12.	Madhya Pradesh	17	23	(-)6	(-)26
13.	Maharashtra	0.1	3	2.9	(-)96
14.	Manipur	Neg	0.5	Neg.	Neg.
15.	Meghalaya	0.8	0.6	0.2	33
16.	Mizoram	-	0	-	-
17.	Nagaland	-	0	-	-
18.	Orissa	9	7	2	29
19.	Punjab	587	870	(-)283	(-)33
20.	Rajasthan	19	NA	NA	NA
21.	Sikkim	-	0	-	-
22.	Tamil Nadu	178	98	80	89
23.	Tripura	2	7	5	71
24.	Uttar Pradesh	1798	1636	162	10
25.	West Bengal	231	373	(-)142	(-)38
26.	Total (1 to 25)	3903	NA		
27.	UTs	2	16	(-) 14	88
28.	All India	3905	NA		
29.	All India excluding Rajasthan	3886	3987	(-)101	(-)3

* to Census figure; Neg.—negligible; NA—not available.

Table 5
Comparative Number of Public/Deep Tubewells Constructed upto 1986-87

S.No.	States	*No. of public tubewells (000)*		*Difference: col.3-col.4*	
		Plan estimate	*Census estimate*	*absolute (000)*	*relative (%)*
1	*2*	*3*	*4*	*5*	*6*
1.	Andhra Pradesh	0.42	13.4	(-)12.98	(-)97
2.	Arunachal Pradesh	-	0	-	-
3.	Assam	0.98	0.3	0.68	227
4.	Bihar	6.04	12.8	(-)6.76	(-)53
5.	Goa	-	Neg	Neg.	Neg.
6.	Gujarat	4.47	22.9	(-)18.43	(-)80
7.	Haryana	1.80	2.7	(-)0.90	33
8.	Himachal Pradesh	0.19	0.1	0.09	90
9.	Jammu & Kashmir	0.15	Neg.	Neg.	Neg.
10.	Karnataka	-	7.1	(-)7.1	(-)100
11.	Kerala	-	0	-	-
12.	Madhya Pradesh	0.40	2.5	(-)2.10	(-)84
13.	Maharashtra	-	4.2	(-)4.2	(-)100
14.	Manipur	-	-	-	-
15.	Meghalaya	-	Neg.	-	-
16.	Mizoram	-	0	-	-
17.	Nagaland	-	0	-	-
18.	Orissa	4.75	2.3	2.45	107
19.	Punjab	1.76	1.5	0.26	17
20.	Rajasthan	0.075	NA	NA	NA
21.	Sikkim	-	0	-	-
22.	Tamil Nadu	-	5.2	(-)5.2	(-)100
23.	Tripura	0.12	Neg.	0.02	20
24.	Uttar Pradesh	24	25.2	(-)1.2	(-)5
25.	West Bengal	3.10	3.1	-	-
26.	Total (1 to 25)	52.44	NA		
27.	UTs	0.25	0.26	Neg.	Neg.
28.	All India	52.69	NA		
29.	All India excluding Rajasthan	52.6	103.8	(-)51.2	(-)49

* to Census figure; Neg.—negligible; NA—not available.

and deep tubewells amounted to about 74 and 21 per cent, respectively, of its total groundwater irrigation potential of the order of 9 mha. The Plan number for shallow tubewells, namely, 1,798 thousands, is found to exceed the corresponding Census number by about 10 per cent. But the Plan estimate of 24 thousand number for deep tubewells happens to be marginally less than the Census one by about (minus) 5 per cent. Giving due weightage to higher capacity of a deep tubewell vis-a-vis shallow one (21:1 as per the Census data which shows that a deep tubewell in U.P. has an irrigation capacity for about 92 crop hectares as against 4.3 crop hectares of a shallow tubewell), the weighted excess in the number of tubewells given in the Plan papers amounts to about 6.5 per cent only. In contrast, the Plan figure of dugwells (1.1 millions) overshoots the Census figure of 0.46 million by over 146 per cent. Assuming a shallow tubewell equals 7 dugwells in U.P., the overall excess of Plan figure on number of wells over Census figure comes to 24 per cent (39 per cent if we do no weighting). Thus, over two-thirds of the difference in the irrigation capacity from two sources is traceable to difference in underlying figures of number of wells in the two sources of data.

The Plan estimate of number of dugwells for U.P. for 1986-87 turns out to be grossly in excess of the Census number because of the procedure of cumulation followed in building up the Plan number, whereby the number of newly-constructed dugwells/ tubewells in year T+1 is simply added on to the number of dugwells/tubewells constructed by the end of year T. Though some retirement or depreciation factors are used in this cumulation exercise, these do not seem to suffice in case premature (before the useful life of a well is over) abandonment of dugwell irrigation occurs. Such abandonment was more common in canal commands during the British times, but in the post-Independence days farmers have tended to abandon dugwell irrigation once they go in for tubewell irrigation of their own. There is little doubt that tubewell 'explosion' of the green revolution period occurred in U.P. also—according to the Census, this state accounted for 1.6 out of about 4 million shallow tubewells.

In the case of Gujarat, the dominant source of groundwater irrigation is dugwell, accounting for nearly 84 per cent of total groundwater irrigation potential as per the breakup available in the

Census. The number of dugwells reported in the Plan document falls short of the Census estimate by about 18 per cent. Interestingly, no difference in respect of shallow tubewells is observable. However, the number of deep tubewells differs sharply, with the Plan number being 80 per cent less than the Census one. Duly reckoning with unit capacity norms for these three types of well irrigation (according to the Census, the unit irrigation capacity for a dugwell, shallow tubewell and deep tubewell in Gujarat averaged to 1.2 ha, 11.6 ha and 24.0 ha, respectively), the overall (weighted) shortfall in the number of wells works out to about 38 per cent (20 per cent if unweighted) as against the 58 per cent shortfall in the case of irrigation potential figure. Thus, much of the difference in the two estimates of groundwater irrigation capacity is due to the difference in number of wells in the two sources of data.

The foregoing analysis clearly suggests that some of the difference in the two estimates of minor irrigation capacity for each state could be due to differences in the underlying norms for unit capacity used for assessing total irrigation capacity from a given number of wells of various types in a state. However, a direct evidence on these differences in unit capacity norms cannot be presented here, because in both the sources these norms are not explicitly spelt out. As for the Census source, the unit norms can be implicitly derived through a simple division of total capacity from a given type of wells, say, shallow tubewells, by the reported number of such tubewells in the Census data. The same derivation, however, could not be carried out for the other source, (*i.e.*, plan figures), in which capacity data are not listed separately by type of wells.

The implied unit capacity norms for dugwells, shallow tubewells and deep tubewells are listed for all states in Table 11. One is immediately struck by the tremendous variability in the magnitude of the (implicit) norm for each of the three major types of wells, namely, dugwell, shallow tubewell and deep tubewell. Thus their mean values at the all-India level, namely, about 1.5 ha for a dugwell, about 3.5 ha for a shallow tubewell and about 74.3 ha for a deep tubewell, need to be viewed with due caution that these vary substantially across states. The source of this variation is partly the differences in well characteristics on the one hand and aquifer parameters on the other, and partly due to subjective or judgemental biases of the assessors of the irrigable area potential of wells, both at the intra-state and the inter-state

Table 6
Comparative Number of Electric Pumpsets installed upto 1986-87

S.No.	States	*No. of electric pumpsets (000)*		*Difference: col.3-col.4*	
		Plan	*Census*	*absolute (000)*	*relative* (%)*
1	*2*	*3*	*4*	*5*	*6*
1.	Andhra Pradesh	831	749	82	11
2.	Arunachal Pradesh	-	0	-	-
3.	Assam	3	2	1	50
4.	Bihar	217	80	137	171
5.	Goa	-	2	(-)2	(-)100
6.	Gujarat	338	190	148	78
7.	Haryana	299	266	33	12
8.	Himachal Pradesh	3	2	1	50
9.	Jammu & Kashmir	2	1	1	100
10.	Karnataka	537	376	161	43
11.	Kerala	161	44	117	266
12.	Madhya Pradesh	553	526	27	5
13.	Maharashtra	1119	869	250	29
14.	Manipur	0.04	-	-	-
15.	Meghalaya	0.06	Neg.	-	-
16.	Mizoram	-	Neg.	-	-
17.	Nagaland	-	-	-	-
18.	Orissa	35	22	13	59
19.	Punjab	491	489	2	Neg
20.	Rajasthan	306	-	-	-
21.	Sikkim	-	-	-	-
22.	Tamil Nadu	1116	797	319	40
23.	Tripura	1	0.4	0.6	150
24.	Uttar Pradesh	570	393	177	45
25.	West Bengal	52	37	15	41
26.	Total (1 to 25)	6625	NA	NA	NA
27.	UTs	32	18	14	78
28.	All India	6657	NA	NA	NA
29.	All India excluding Rajasthan	6351	4861	1490	31

* to Census figure; Neg.—negligible; NA—not available.

Table 7
Comparison of Plan and Census Estimates of Total Minor Irrigation Utilisation by State, 1986-87

S.No.	*State*	*Plan estimate (000 ha)*	*Census estimate* (000 ha)*	*Difference: Col.3-Col.4*	
				absolute (000 ha)	*relative (%)*
1	*2*	*3*	*4*	*5*	*6*
1.	Andhra Pradesh	2308	2726	(-)418	(-)15
2.	Arunachal Pradesh	43	43	0	0
3.	Assam	385	160	225	141
4.	Bihar	3445	2879	566	20
5.	Goa	14	12	2	17
6.	Gujarat	1659	1439	220	15
7.	Haryana	1394	1743	(-)349	(-)20
8.	Himachal Pradesh	107	166	(-)59	(-)35
9.	J & K	335	NA	(-)NA	NA
10.	Karnataka	1159	1092	67	6
11.	Kerala	394	313	81	26
12.	M.P.	2046	2061	(-)15	(-)1
13.	Maharashtra	1880	2957	(-)1077	(-)36
14.	Manipur	36	14	22	157
15.	Meghalaya	34	35	(-)1	(-)3
16.	Mizoram	7	6	1	17
17.	Nagaland	50	51	(-)1	(-)2
18.	Orissa	1054	475	579	122
19.	Punjab	3172	4579	(-)1407	(-)31
20.	Rajasthan	2011	NA	NA	NA
21.	Sikkim	12	16	(-)4	(-)25
22.	Tamil Nadu	1998	2116	(-)118	(-)6
23.	Tripura	59	29	30	103
24.	U.P.	12589	8677	3912	45
25.	West Bengal	1663	1860	(-)197	(-)11
26.	Total (1 to 25)	37853	33348	4505	14
27.	UTs	74	97	(-)23	(-)24
28.	All India	37927	NA	NA	NA
29.	All India excluding Rajasthan	35916	33445	2421	7

* Derived from Table 1F, p.84. Utilisation figures are exclusive of area benefitting from supplementary irrigation. NA—not available.

Table 8

Statewise Comparison of Plan and Census Estimates of Minor Irrigation Utilisation by Groundwater and Surface Water

S.No.	State	Ground Water Irrigation				Surface Water Irrigation			
		Plan estimate (000ha)	Census estimate* (000ha)	Difference : Col.3-Col.4 absolute (000ha)	Difference : Col.3-Col.4 relative (%)	Plan estimate (000ha)	Census estimate* (000ha)	Difference: Col.7-Col.8 absolute (000ha)	Difference: Col.7-Col.8 relative (%)
1	*2*	*3*	*4*	*5*	*6*	*7*	*8*	*9*	*10*
1.	Andhra Pradesh	1274	1593	(-)319	(-)20	1033	1133	(-)100	(-)9
2.	Arunachal Pradesh	-	-	-	-	43	43	0	0
3.	Assam	82	32	50	156	303	127	176	139
4.	Bihar	2309	2126	183	9	1136	753	383	51
5.	Goa	0.2	2	(-)1.8	(-)90	14	10	4	40
6.	Gujarat	1516	1418	(-)98	7	143	21	122	581
7.	Haryana	1350	1734	(-)384	(-)22	34	9	25	278
8.	Himachal Pradesh	8	15	(-)7	(-)47	99	151	(-)52	(-)34
9.	J & K	6	6	-	-	329	NA	NA	NA
10.	Karnataka	495	601	(-)106	(-)18	664	490	174	36
11.	Kerala	57	55	2	(-)4	336	258	78	30
12.	M.P.	1231	1463	(-)232	(-)16	772	598	174	29
13.	Maharashtra	1262	2246	(-)984	(-)44	618	711	93	(-)13
14.	Manipur	-	-	-	-	36	14	22	157
15.	Meghalaya	9	5	4	80	25	30	(-)5	(-)17
16.	Mizoram	-	0	-	-	7	6	1	17
17.	Nagaland	0.2	Neg.	Neg.	Neg.	50	51	(-)1	(-)2

(Contd...)

18.	Orissa	493	112	381	340	562	363	199	55
19.	Punjab	3137	4574	(-)1437	(-)31	35	5	30	600
20.	Rajasthan	1646	NA	NA	NA	365	NA	NA	NA
21.	Sikkim	-	0	-	-	12	16	(-)4	(-)25
22.	Tamil Nadu	1163	1278	(-)115	(-)9	835	839	(-)4	Neg.
23.	Tripura	9	6	3	50	50	23	27	117
24.	U.P.	11723	8457	3266	39	866	221	645	292
25.	West Bengal	648	991	(-)343	(-)35	1015	769	246	32
26.	Total (1 to 25)	28429	26714	1715	6	9382	6640	2742	41
27.	UTs	60	76	(-)16	(-)21	15	21	(-)6	(-)29
28.	All India	28490	NA	NA	NA	9438	NA	NA	NA
29.	All-India excluding Rajasthan	26844	26790	(-)54	Neg.	9073	6661	2412	36

* Compiled from Table 3F, p. 86. Utilisation figures are exclusive of area benefitting from supplementary irrigation.

levels. These biases may lie hidden in the tacit assumptions made by an assessor in respect of crucial determinants like availability of electricity for pumpset operation and crop mix likely to be adopted by well owners.

Whereas the Plan norm for irrigation potential for a well is mainly the handiwork of state officials closely involved with minor irrigation development in a state, the same parameter in the Census was supposed to be assessed by the well owner. To do so, must have been a heroic task for many well owners. While doing so, he would have been truly baffled as to what ought to go into the determination of irrigation potential of a well. On the 'supply' side, he would need to know how much volume of water yield he can expect from his well at different points of time in a year, notably during the three crop seasons of kharif, rabi and summer. On the 'demand' side, he would need to reckon with the size of his land holding and its fragmentation in space, as this would act as a binding constraint on irrigation capacity in case he happens to own a copious well—this constraint is loosened if groundwater market exists for his surplus well water.[5] Then, crop prices and soil types would influence his crop-mix: heavily-irrigated crops and crops occupying the field for long duration (*e.g.,* sugarcane), if chosen would ceteris paribus reduce irrigation capacity of a well. Therefore, one cannot affirm that the Census figures in respect of minor irrigation potential are *necessarily* indicative of ground truth in this regard, while the corresponding Plan estimate is purely a normative number of questionable reality. Probably, both the Census and the Plan numbers in respect of irrigation potential are deficient in realism as the very concept of irrigation potential is not that well defined or definable. If we wish to improve our data base on minor irrigation potential, we better not take seriously farmers' response as to what amount of irrigation capacity (in crop area terms) they possess. Instead, average crop area irrigated per well, as reported by them, may be used as a realistic measure of capacity.[6]

---

5. Some tubewell owners in Gujarat install tubewells as an agri-business proposition, *i.e.* to earn income from sale of tubewell water. Sample surveys of tubewells in this state, known for its groundwater markets, do reveal high percentage of water selling as compared to onfarm use of water by tubewell-owning farmers. This may explain why irrigation capacity per shallow tubewell for this state turns out to be comparatively high (see Table 11).
6. If in a census year, drought (or some other transitory factor) leads to low level of capacity utilisation, one would then be justified in adopting the corresponding figure of farmer's own assessment of capacity.

II. MINOR IRRIGATION WORKS NOT IN USE

One useful information that has been gathered through the Census of minor irrigation is about minor irrigation works not found in use during the reference period of the Census *viz.* 1986-87. This much-needed information was sought on the presumption that small-scale irrigation works become defunct or unusable for supply side reasons: siltation of tank beds and breaches of tank bunds in the case of tank irrigation works; in the case of wells the non-use could be associated with following problems:

1. subsidence or collapse of structure;
2. burnt-out electrical systems of wells equipped with electric motors;
3. damaged/worn-out pumpset system;
4. choked filters in the case of tubewells;
5. caved-in cavities in the case of cavity tubewells.

The incidence of an irrigation work remaining out of use is specific to small-scale irrigation units. According to the Census of minor irrigation, about 1.6 million such works were not used for irrigation during 1986-87. This number constitutes nearly 13 per cent of about 11.7 million minor irrigation works in existence at that time. The highest incidence of non-use (19 per cent) is found in the case of dugwells which have the lowest unit irrigation capacity of 1.5 crop ha per dugwell among all the five categories of minor irrigation. But then the lowest incidence of non-use occurs not in sources with comparatively high unit irrigation capacities, namely, deep tubewell (unit capacity of about 74.3 ha per tubewell) and surface flow irrigation shcemes (unit capacity = about 13.8 ha), but in shallow tubewells (unit capacity = 3.5 ha) and surface lift irrigation schemes (unit capacity = 5.0 ha). The comparative data for the five categories of minor irrigation works is given below.

		dug-wells	*surface flow irri-gation*	*deep tube-wells*	*shallow tube-wells*	*surface lifts*
1.	Incidence of works not in use (%)	19	11	8	1	2
2.	Average irrigation potential per work (ha)	1.5	13.8	74.3	3.5	5.0

Table 9

Per Cent Small-scale Irrigation Works Not in Use during 1986-87

	State	*Dugwells*	*Shallow tubewells*	*Deep tube-wells*	*Surface flow irri-gation*	*Surface lift irri-gation*
1.	Andhra Pradesh	11	2	13	13	3
2.	Arunachal Pradesh	x	x	x	7	50
3.	Assam	0	10	10	Neg	7
4.	Bihar	6	10	10	4	12
5.	Goa	6	0	0	2	2
6.	Gujarat	55	14	6	63	20
7.	Haryana	22	Neg	2	0	0
8.	Himachal Pradesh	7	1	2	3	6
9.	J & K	2	9	3	Neg	7
10.	Karnataka	11	5	6	31	2
11.	Kerala	NA	x	x	9	1
12.	M.P.	14	4	5	29	Neg
13.	Maharashtra	16	11	18	4	2
14.	Manipur	x	17	x	-	-
15.	Meghalaya	0	34	0	2	3
16.	Mizoram	x	x	x	9	0
17.	Nagaland	0	x	x	3	x
18.	Orissa	12	16	19	9	12
19.	Punjab	67	0	2	0	0
20.	Rajasthan	NA	NA	NA	NA	NA
21.	Sikkim	x	x	x	12	x
22.	Tamil Nadu	10	3	5	2	Neg
23.	Tripura	11	8	7	6	12
24.	U.P.	47	1	5	26	3
25.	West Bengal	4	2	5	2	2
26.	UTs	5	10	3	1	0
27.	All India excluding Rajasthan	19	1	8	11	2

Notes: 'x' means this class of works do not exist in this state.
Neg. means negligible (less than one half).

Table 10
Variations in the Incidence of Non-Use of Minor Irrigation Works

Category of MI work	*High incidence of non-use*	*Low incidence of non-use*
I. Dugwells	1. Punjab(67)	1. J & K(2)
	2. U.P.(47)	2. W. Bengal(4)
	3. Gujarat(55)	
	4. Haryana(22)	
II. Shallow Tubewells	1. Gujarat(14)	1. Haryana(neg)
	2. Maharashtra(11)	2. J & K(9)
	3. Orissa(12)	3. Punjab(0)
		4. U.P.(1)
		5. W. Bengal(2)
		6. Tamil Nadu(3)
		7. Andhra Pradesh(2)
		8. M.P.(4)
III. Deep Tubewells	1. Maharashtra(18)	1. Goa(0)
	2. Andhra Pradesh(13)	2. Haryana(2)
	3. Orissa(19)	3. H.P.(2)
		4. J & K(3)
		5. U.P.(5)
IV. Surface Flow Irrigation		
	1. Gujarat(63)	1. Punjab(0)
	2. U.P.(26)	2. Haryana(0)
	3. Karnataka(31)	3. J & K(neg)
	4. M.P.(24)	4. W. Bengal(2)
	5. Andhra Pradesh(13)	5. Tamil Nadu(2)
V. Surface Lift Irrigation		
	1. Arunachal Pradesh(50)	1. A. Pradesh(3)
	2. Gujarat(20)	2. Karnataka(2)
	3. Orissa(12)	3. Kerala(1)
	4. Tripura(12)	4. Maharashtra(2)
		5. M.P.(neg)

Neg.—negligible.

The extent of spatial variation in the incidence of non-use for any category of works is indicated in Table 9. The gravest problem of non-use of minor irrigation is reported from Gujarat. As between types of irrigation, the range for dugwells and surface flow irrigation schemes is much more than for tubewells and surfacewater lift irrigation schemes. Select states reporting unusually high/low non-use of any category of irrigation works are listed in Table 10 alongwith the percentage of non-use within parentheses.

What are the causes behind nearly 13 per cent of minor irrigation works being not in use during the census year 1986-87? Since there is no information in this regard in the Census report, we cannot ascertain as to what extent the phenomenon of disuse was due to transitory factors, like (1) a well/tank drying up because of severe drought and (2) a farm being not cultivated at all in the census year for reasons of serious sickness/injury to its owner. It is well known that the transitory factor (1) *viz.* severe drought, does render shallow wells and tanks with small catchments incapable of supporting irrigated farming. Admittedly, the Census year 1986-87 cannot be characterised as a drought year for the nation as a whole. Yet, there could be pockets of drought even in a year of above-normal rainfall in this vast country.

In point of fact, the extremely high incidence of works being out of use in the state of Gujarat for the Census year is primarily due to drought in that state. Since this drought was preceded by a drought in 1985-86 and below normal rains during 1984-85, the severe drought of 1986-87 led to extensive non-availability of water in wells and small storage works in the state. No wonder, 467 thousand minor irrigation structures (mostly dugwells, the mainstay of minor irrigation in the state) were not found in use during 1986-87. Thus if Gujarat's works are excluded from the national aggregate analysis, a marked reduction in the incidence of non-use of minor irrigation structures is found (from 13 to 10 per cent). This revised estimate of 10 per cent for the incidence of non-use of minor irrigation works might be taken as a crude measure of the extent to which minor irrigation works had gone out of use for ever due to non-transitory factors (overage, damage beyond normal repair, prematurely abandoned because farmers got access to some other source of irrigation etc.). On this reckoning, one might say that one-tenth of our created minor irrigation potential which is anticipated to be achieved by the end of the Eighth Five Year Plan (*i.e.* by 31st March, 1997) need to be *derated* by a magnitude of about 5.784 million hectares.

It needs to be pointed out here that our estimate of minor irrigation capacity having fallen into disuse is on the high side when compared to that mentioned in the report of the Working Group on Minor Irrigation for IXth Plan. On page 24 of this report it is mentioned that about 3.612 mha of minor irrigation potential had 'gone out of use'. Two factors might account for this disagreement by a margin of 2 mha. First, we have applied the 10 per cent factor of deration to (likely) capacity in minor irrigation as on 31st March, 1997 (*i.e.,* the end of Eighth Five Year Plan period), and not to the capacity in existence at the time of the Census of minor irrigation. In other words, we derate an achievement level of 57.84 mha in created minor irrigation capacity whereas the Minor Irrigation Group seems to have derated the level of about 39 mha achieved by 1986-87. Second, the derating factor used by the Group is slightly lower than one-tenth (probably around one-eleventh).

Our estimate of minor irrigation capacity gone out of use needs to be firmed up in the following manner. First, the exercise should be done at state level[7]--it cannot be done at district level because the Plan data on created irrigation potential is compiled at state level. Second, attempt should be made to do the exercise for each of the five categories of minor irrigation.

III. SMALLNESS OF A MINOR IRRIGATION WORK

The phrase 'minor irrigation' connotes smallness of an irrigation work both in terms of investment required to establish such a work and irrigation benefits expressed in irrigated area figures. The Census of minor irrigation provides necessary data for sizing up the small-scaleness of various categories of minor irrigation works in benefited area terms for each district of India, as also for each state

7. This cannot be done for Rajasthan where the Census was not conducted, nor it should be done for Gujarat where the drought put many works out of use.

Table 11
Unit Created Irrigation Potential from Small Scale Irrigation Works, 1986-87

(in ha)

S. No.	*State*	*Dugwells*	*Shallow tube-wells*	*Deep tube-wells*	*Surface flow irriga-tion*	*Surface lift irriga-tion*
1.	Andhra Pradesh	1.2	2.4	4.9	18.3	7.5
2.	Arunachal Pradesh	X	X	X	13.2	17.1
3.	Assam	0.4	2.5	44.8	33.3	38.5
4.	Bihar	1.5	3.5	33.7	14.2	36.7
5.	Goa	0.6	0.8	7.0	3.4	5.7
6.	Gujarat	1.2	11.6	24.0	182.0	21.1
7.	Haryana	2.2	5.1	14.7	37.7	76.3
8.	Himachal Pradesh	2.2	6.0	38.7	9.6	23.2
9.	J & K	0.8	2.1	48.3	NA	105.6
10.	Karnataka	1.2	2.4	2.5	19.7	7.3
11.	Kerala	0.4	X	X	19.2	3.1
12.	M.P.	1.4	4.3	9.2	17.1	3.7
13.	Maharashtra	1.6	2.2	3.5	28.4	5.2
14.	Manipur	X	X	X	118.4	200.0
15.	Meghalaya	6.3	16.5	53.0	3.2	116.3
16.	Mizoram	-	X	X	1.6	88.3
17.	Nagaland	0.3	X	X	2.5	X
18.	Orissa	0.5	8.3	30.0	30.5	21.8
19.	Punjab	1.2	5.4	50.7	19.0	11.3
20.	Rajasthan	NA	NA	NA	NA	NA
21.	Sikkim	X	X	X	30.5	X
22.	Tamil Nadu	1.1	1.7	2.7	26.8	3.6
23.	Tripura	0.1	0.6	56.3	6.0	17.5
24.	U.P.	0.9	4.3	92.3	4.1	3.9
25.	West Bengal	0.7	2.6	63.3	6.0	3.4
26.	UTs	2.1	5.1	11.4	16.3	9.1
27.	All India excluding Rajasthan	1.5	3.5	74.3	13.8	5.0

NA—not available.
X—works not in existence in this State.

of the country. The same cannot be easily generated from existing official statistics on number of minor irrigation structures and area under minor irrigation sources like tanks, wells, etc.

In assessing the smallness, we have to divide reported benefited area by the corresponding number of structures. To illustrate, the smallness of a dugwell can be sized up by dividing total crop area irrigated by dugwells(A) by number of dugwells(N). One problem we face here is that of supplementary irrigation. Another problem is that of the resultant estimate being unduly depressed in a district because of some transitory factor such as drought or acute power shortage faced by farmers in the Census year. In view of these difficulties the true size of an irrigation work is better measured if 'A' is replaced by irrigation capacity in the above derivation.

Measurement of capacity, whether of an economic activity or an individual, is a daunting task.[1] Because of the administrators'/researchers' interest in assessing the level of capacity utilisation, valiant attempts are made by empiricists to come to grips with this task. While we would not go into how minor irrigation capacity is being assessed, we would look at the results of such an assessment as reported in the Census of minor irrigation. The unit irrigation capacity for five categories of minor irrigation works is derived from a simple division of the reported total irrigation potential created under each category by 1986-87 by the corresponding number of irrigation works of a given category in existence.

Minor irrigation works, unlike major and medium irrigation works, are, by and large, well suited to individual ownership by Indian farmers who have predominantly small land holdings. This is because of the perception that their unit capacity is low enough to be utilised economically even on the small holdings. This is most true for dugwells whose average irrigation capacity works out to nearly 1.5 crop hectares of area in a whole agricultural year. To a large extent, this is also true for shallow tubewells for which the average capacity works out to nearly 3.5 ha of crop area. To a much lesser extent it is true for surface lift irrigation schemes. But deep tubewells and surface flow irrigation schemes have usually much more irrigation capacity than the one that can be fully utilised by an individual farmer on his own farm. According to the

Census statistics, the unit irrigation capacity averaged about 74 ha for a deep tubewell, and about 14 ha for a surface water based gravity scheme. This explains why these two categories of irrigation works are predominantly under the public and community aegis in India.

Statewise unit irrigation capacities for the five categories of minor irrigation works, listed in Table 11, reveal great variations across states for the same category of work. For a dugwell, the inter-state range in unit capacity is 0.1-6.3 ha. For a shallow tubewell, this range is 0.6-16.5 ha. The range is 2.5 to 92.3 ha for a deep tubewell, 1.5-182 ha for a gravity flow scheme, and 3.1-200.0 ha for lift irrigation scheme. Such large variations for each category of work arise because of a variety of reasons into which we need not go presently. Let us see how the states tend to cluster in respect of unit irrigation capacity for each category of work.

Irrigation Capacity	**Dugwells**
Less than 0.5 ha	Nagaland (0.3), Tripura (0.1)
0.5-1.0ha	Goa(0.6),J& K(0.8), Orissa(0.5), U.P.(0.9), W. Bengal (0.7)
1.1-1.5 ha	Andhra(1.2), Karnataka(1.2), M.P.(1.4), Punjab(1.2), Tamil Nadu(1.1) Gujarat(1.2)
1.6-2.0 ha	Maharashtra (1.6)
2.1-3.0 ha	Haryana(2.2), UTs(2.1)
	Shallow Tubewells
Less than 1.0 ha	Goa(0.8), Tripura(0.6).
1-2 ha	Maharashtra(2.2), J & K(2.1), Tamil Nadu(1.7)
2-3 ha	Andhra(2.4), Karnataka(2.4), W. Bengal(2.6)
3-5 ha	M.P.(4.3), U.P.(4.3), Haryana(5.1)
Above 5 ha	Gujarat(11.6), H.P.(6.0), Orissa(8.3), Punjab(5.4), UTs(5.1)

Deep Tubewells

Less than 5 ha	Karnataka(2.5), Maharashtra(3.5), Tamil Nadu(2.7), Andhra(4.9)
5-15 ha	Goa(7.0), Haryana(14.7), M.P.(9.2), UTs(11.4)
15-25 ha	Gujarat(24.0)
25-50 ha	H.P.(38.7), J&K(48.3),Orissa(30.0), Tripura(56.3)
50-75 ha	Punjab(50.7), W.Bengal(63.3)
Above 75 ha	U.P.(92.3)

Surface Flow Irrigation

Less than 10 ha	Goa(3.4), H.P.(9.6), Mizoram(1.5), Nagaland(2.5), U.P.(4.1), W.Bengal(6.0), Tripura(6.0)
10-25 ha	Andhra(18.3), Arunachal(13.2), Karnataka(19.7), Kerala(19.2), M.P.(17.1) UTs(16.3), Maharashtra(17.1), Punjab(19.0)
25-50 ha	Haryana(37.7), Tamil Nadu(26.8) Orissa(30.5), Sikkim(30.5)
Above 50 ha	Gujarat(182.8)

Surface Lift Irrigation

Less than 5 ha	Kerala(3.1),M.P.(3.7),Tamil Nadu(3.6), U.P.(3.9)
5-10 ha	Andhra(7.5),Goa(5.7), Karnataka(7.3), Maharashtra(5.2), UTs(9.1)
10-25ha	Arunachal(17.1),Gujarat(5.7), H.P.(23.2), Orissa(21.8), Tripura(17.5)
Above 25 ha	Haryana (76.3), J & K(105.6)

It will be instructive to assess how the unit irrigation capacity for a given type of structure, say, a dugwell varies across districts. This is, however, not attempted in the present study.

IV. ELECTRICITY CONSTRAINT IN PUMPSET OPERATION

Electric pumpsets have now become the single most important waterlifting mechanism in Indian minor irrigation. The Census statistics reveal that of the nearly 7.7 million dugwells nearly 3.20 millions had electric pumpsets (diesel pumpsets numbered over one million). Likewise, of the about 4 million shallow tubewells, about 1.4 million were fitted with electric motors, the rest being mostly diesel driven. In the case of surface waterlifts, the number of electric ones was 0.18 million as compared to 0.21 million diesel ones. Though separate breakup for deep tubewells is not available, we know otherwise that most of them are equipped with electric motors: as many as 74,953 out of total 102,054 deep tubewells were fitted with submersible pumps which are invariably electrically driven.

Inefficiencies in the electricity supply system, which is exclusively under state control in all the states, seriously impinge on the performance of agriculture based on minor irrigation works. While voltage fluctuations and non-supply of power on all three phases simultaneously give rise to functional problems, power sheddings as such hinder timely completion of irrigation operation. In particular, night time instead of day-time supply of power adversely affects uniform application of water to crops, thereby lowering the efficiency of water use. The reference year of 1986-87 is marked by severe power shedding for pumpsets. Owners of pumpsets reported varying durations for which electricity was available for pumping in different states during *peak* season. Categorywise and statewise picture of daily power availability during the Census year is shown in Table 12. On the average, owners of shallow tubewells had peak time power availability for just 6 hours per day (of 24 hours), being as low as 3 hours in Tripura and as high as 18 hours in Goa. In the case of deep tubewells, which are largely in the public sector and many with dedicated power feeders, the availability of power supply was barely 3 hours per day with a range of 2 to 10 hours across states. For lift irrigation, the supply position was somewhat better, about 4 hours per day. This better availability in comparison to tubewells arises because of a regional element: while lift irrigation is concentrated in states with lesser power shortages, tubewells are concentrated in states with more power shortages.

For all states and all categories of minor irrigation taken together, the daily peak time availability of power averaged to 4 hours. This shortage of power must have affected adversely the rate of utilisation of capacity, especially of government tubewells

Table 12
Daily Availability of Electricity for Rural Pumpsets during Peak Season, 1986-87

S.No.	*State*	*Dugwells*	*Shallow tubewells*	*Deep tubewells*	*Surface Lift irrigation*	*State average*
1.	Andhra Pradesh	6	6	8	7	6
2.	Arunachal Pradesh	NA	NA	NA	NA	NA
3.	Assam	NA	NA	NA	NA	NA
4.	Bihar	4	4	4	5	4
5.	Goa	18	18	8	13	13
6.	Gujarat	6	6	6	4	5
7.	Haryana	8	8	8	9	8
8.	Himachal Pradesh	5	5	8	10	6
9.	J & K	NA	NA	3	NA	1
10.	Karnataka	8	8	8	8	8
11.	Kerala	NA	NA	NA	NA	NA
12.	M.P.	10	10	10	12	11
13.	Maharashtra	2	2	10	7	4
14.	Manipur	NA	NA	NA	NA	NA
15.	Meghalaya	NA	NA	NA	4	1
16.	Mizoram	NA	NA	NA	2	1
17.	Nagaland	NA	NA	NA	NA	NA
18.	Orissa	8	8	8	8	8
19.	Punjab	7	7	10	4	7
20.	Rajasthan	NA	NA	NA	NA	NA
21.	Sikkim	NA	NA	NA	NA	NA
22.	Tamil Nadu	6	6	2	5	4
23.	Tripura	3	3	5	5	4
24.	U.P.	7	7	8	7	7
25.	West Bengal	7	7	9	6	7
26.	All India excluding Rajasthan	6	6	3	4	4

and surface lift irrigation schemes which need to be operated, more or less, continuously during the peak season.[8] This underlines the imperative of making commensurate investments in rural power supply when planning for expansion of minor irrigation capacity. Unfortunately, champions of minor irrigation cause do not fully appreciate such power-related investment costs of minor irrigation. Not reckoning with these costs, they continue to view minor irrigation as a low-cost option for irrigation development in India.

V. UTILISATION OF MINOR IRRIGATION POTENTIAL

Utilisation of irrigation capacity has two aspects which need to be distinguished. One pertains to *absolute* utilisation, and the other to *rate* of utilisation. Absolute utilisation is a measure of crop area benefited by irrigation. This area may be viewed in different dimensions: (1) by type of crop or crop groups (*e.g.,* foodgrains and non-foodgrains area); (2) by crop season such as rabi, kharif, summer area, and (3) by solo (single irrigation source) or joint (multiple irrigation sources). The Census of minor irrigation offers information in respect of the dimensions (2) and (3), something which is just not available for most states in the LUS records. The only possible cross-check type of analysis that can be carried out with some degree of confidence is one of comparing the Census estimate of *total* crop area irrigated by minor irrigation sources in 1986-87 with the corresponding estimate of minor irrigation utilisation arrived at by the Planning Commission. Such a comparison is conducted in Tables 7 and 8.

The Plan estimate for minor irrigation utilisation for the nation as a whole for 1986-87 is not markedly different from that of the Census. Admittedly, the former exceeds the latter by a margin of 2.4 m ha, but in relative terms by just 7 per cent (vide Table 7). Moreover, this difference is mitigated by the fact that as per the Census results minor irrigation works provided *supplementary* irrigation to about 4 million hectares. Assuming that (a) a minor irrigation source contributes as much water in supplementary irrigation as a major source of irrigation and (b) one

8. Power shortages in Uttar Pradesh are believed to have *ceteris paribus* reduced the annual duration of working of state government tubewells by about 1000 hours per year, and are said to be one major factor behind the decline in the running hours per tubewell over time.

Table 13
Percentage Utilisation of Irrigation Capacity

S.No.	*State*	*Dugwells*	*Shallow Tube-wells*	*Deep tube-wells*	*Surface flow Irriga-tion*	*Surface lift Irriga-tion*	*All Works*
1.	Andhra Pradesh	86	92	82	72	75	81
2.	Arunachal Pradesh	X	X	X	71	98	71
3.	Assam	65	61	65	77	57	66
4.	Bihar	90	91	78	105	84	92
5.	Goa	56	75	100	72	64	67
6.	Gujarat	33	67	72	7	25	37
7.	Haryana	99	89	91	97	205	90
8.	Himachal Pradesh	83	76	72	88	55	85
9.	J & K	116	116	100	NA	NA	NA
10.	Karnataka	80	91	88	43	70	63
11.	Kerala	100	X	X	76	72	79
12.	M.P.	87	100	95	77	92	86
13.	Maharashtra	97	100	92	70	96	93
14.	Manipur	X	X	X	68	60	66
15.	Meghalaya	90	39	100	80	58	70
16.	Mizoram	X	X	X	90	98	90
17.	Nagaland	100	X	X	97	X	97
18.	Orissa	58	43	34	70	40	57
19.	Punjab	92	96	77	57	68	96
20.	Rajasthan	NA	NA	NA	NA	NA	NA
21.	Sikkim	X	X	X	100	X	100
22.	Tamil Nadu	67	95	105	72	70	71
23.	Tripura	96	100	35	69	39	49
24.	U.P.	76	92	71	71	80	86
25.	West Bengal	72	87	46	74	65	75
26.	UTs	94	100	96	88	95	96
27.	All India	71	93	72	77	74	83

minor irrigation source is not used in conjunction with another minor irrigation source (a strong assumption for southern India where groundwater is a supplementary source of irrigation within tank commands), one can say Indian planners have not exaggerated the ground reality while making estimates of capacity utilisation in respect of minor irrigation. However, this is a valid statement in the aggregate and cannot be deemed applicable at state level, because the ground reality is markedly at variance with planners' position in several states (vide Table 7). For instance, the planners' estimate exaggerates ground reality by 122 per cent in Orissa, 141 per cent in Assam and 45 per cent in U.P. But substantial underassessment of ground reality emerges for Maharashtra (-36 per cent) and Punjab (-31 per cent).

Groundwater and surface water utilisation of minor irrigation capacity are separately compared in Table 8. At the national level, the problem of Plan estimate of capacity utilisation exceeding the Census estimate is almost wholly traceable to surface water sources, because there emerges a negligible difference in respect of groundwater utilisation. More precisely, the Plan estimate of capacity utilisation in the case of surface water minor irrigation capacity exceeded the Census estimate by nearly 36 per cent. Barring Andhra Pradesh, Himachal Pradesh, Meghalaya, Sikkim and UTs, most other states fell in this pattern, namely, the Plan figure for utilisation for surface minor irrigation turning out in excess of the Census figure.

Let us now examine the Census results in respect of rate of capacity utilisation for minor irrigation works. That is to say, what was the percentage of minor irrigation capacity (said to have been established as per the respondents in the Census) that was utilised in 1986-87. The legend that all the irrigation capacity developed through minor irrigation works, unlike through major irrigation works, is fully utilised is belied by the Census.[9] At the all-India level, the rate of capacity utilisation was about 80 per cent, not 100 per cent. With the solitary exception of the state of Sikkim, 100 per cent utilisation is not found in any other state (vide Table

9. To some extent the Planning Commission is a party to the perpetuation of this legend, because plan documents continued to show full capacity utilisation against minor irrigation head till 1980-81 when it began showing some underutilisation under this head also.

13), nor for that matter for any of the five categories of minor irrigation works--we shall discuss later categorywise utilisation. Some salient results of statewise and categorywise analysis are discussed in what follows.

STATEWISE PICTURE

The worst or lowest rate of minor irrigation capacity is in Gujarat. Here, there was 37 per cent capacity utilisation, and the rate improves to 38 per cent when supplementary irrigated area is reckoned with. As already pointed out (while discussing minor irrigation works not found in use), the severe drought could be a primary factor behind such low level of utilisation.[10] No wonder, the rate of capacity utilisation was hardly 7 per cent in the case of surface water flow irrigation works, 25 per cent for surface water lift irrigation works, and 33 per cent for dugwells—all these three sources are proverbially known for their vulnerability to drought. This evidence, besides falsifying the legend of full capacity utilisation, underlines the poor protective role of minor irrigation works during drought time in low rainfall regions.[11]

Tubewells are usually not vulnerable to drought. However, if a drought is preceded by a drought, water availability in them can be adversely affected. This is why the rate of capacity utilisation was 67 per cent for shallow tubewells—a bit better (72 per cent) for deep tubewells. Even if we reckon with supplementary irrigated area, at least one-fourth of the tubewell capacity remained unutilized. Thus, even a dependable source of irrigation like tubewell can perform well below par when two or more droughts come in a row.[12]

Orissa too reveals substantial underutilisation of minor irrigation capacity. The state is reported to have a created irrigation potential

10. Drought of 1986 was widespread in the state of Gujarat: Gujarat meteorological division recorded 407 mm (N = 846 mm) and Saurashtra and Kutch meterological division 326 mm (N = 609 mm). Rainfall was below normal in the preceding two years also *e.g.* in Gujarat division it was 538 and 751 mm during 1985 and 1984, respectively, and in the other division it was 326 and 478 mm, respectively.

11. Comparatively much less vulnerability to drought is found in the case of major irrigation works, especially when these are backed by big reservoirs. In Gujarat, net area under canal irrigation did shrink by 27 percent in 1986-87.

12. Here, performance is being viewed in crop area terms only. A better measure of performance would be in crop production terms.

of the order of 800 thousand ha, almost half of it from surface flow irrigation schemes. The rate of utilisation averages to 57% (59% if supplementary irrigated area is included) for the whole minor irrigation segment. The problem of underutilisation exists in all categories of sources. There are indeed many states which reveal above 90 per cent capacity utilisation in minor irrigation: Bihar, Haryana, Maharashtra, Mizoram, Nagaland, Punjab and Sikkim (also Union Territories as a group). Yet, circumspection is needed in accepting this rosy picture. This caution is all the more necessary where state governments have ownership of minor irrigation works. Such ownership is predominant in the case of deep tubewells, and considerable in the case of surface flow and surface lift irrigation schemes.

VI. CATEGORYWISE MINOR IRRIGATION PICTURE

In this concluding section we propose to review some salient results of the Census of Minor Irrigation (CMI). As already pointed out, a minor irrigation source in our country is predominantly used in solo, that is, singly. Its use in conjunction with any other source of irrigation, defined as supplementary irrigation in the CMI, accounted for about one-tenth of its total utilisation during 1986-87. According to data in Table 14, nearly 4 million hectare area benefited from supplementary irrigation during 1986-87, and this constituted 10.7 per cent of total area benefitted by minor irrigation sources. This shows that the phenomenon of much-desired conjunctive irrigation is still in its early phase in India.

The productive value of irrigation water is greater (a) if such water is available in dry seasons of an agricultural year, more so during the rabi season and (b) if irrigation is available throughout the year for raising high-value, annual or perennial crops. On the other hand, irrigation to crops during the wet season tends to be largely protective against rainfall failure, especially in high rainfall areas. Therefore, seasonal orientation of an irrigation work is of considerable importance, because it provides a measure of irrigation resource being productive or protective in character. Seasonality in utilisation by irrigation category is highlighted in Table 15. The CMI data reveals that almost half the area (47 per cent) benefited by minor irrigation works during 1986-87 was for crops of the rabi season. Irrigation for kharif crops accounted for two-fifths of total irrigation. It is only in the case of flow irrigation works based on surface water that kharif irrigation accounts for 65 per cent of total irrigated area. That throws some light on why the productive impact of such works, especially tanks, turns out to be lower vis-a-vis well or lift irrigation.

Table 14
Typewise Minor Irrigation Capacity and Its Utilisation, 1986-87

S. No.	*Type*	*Created capacity (000 ha)*	*Utilisation (000 ha)*			*Rate of capacity utilisation (%)*
			sole irrigation	*supplementary irrigation*	*total*	
(1)	*(2)*	*(3)*	*(4)*	*(5)*	*(6)*	*(7)*
1.	Shallow tubewells	16,863	15,612	2,366	17,978	93-107
2.	Deep Tubewells	3,858	2,764	380	3,144	72-81
3.	Dugwells	11,778	8,414	614	9,028	71-77
4.	Groundwater (1to3)	32,493	26,784	3,360	30,150	82-93
5.	Surface water:					
	(i) flow	6,342	4,890	359	5,244	77-83
	(ii) lift	2,449	1,802	311	2,113	74-86
	Total	8,790	6,662	670	7,362	76-84
	All-India[1] (1to5)	41,284	34,445	4,030	37,512	83-91

1. Excludes Rajasthan. The excluded state accounted for about 5 per cent of the all-India minor irrigation potential as per the estimates of minor irrigation capacity created by the mid-eighties and mentioned in our Plan reports.

The aggregative picture of capacity utilisation in the minor irrigation segment, as already mentioned in the preceding section V, is rather reassuring, as somewhere between 83 and 91 per cent of the irrigation capacity was utilised. The rate of capacity utilisation for groundwater sources, taken together, was perceptibly better (82-93 per cent) than for surface water irrigation (76-84 per cent). For each of the five types of minor irrigation works, the picture of capacity utilisation is discussed separately in what follows.

Table 15
Seasonwise Utilisation of Minor Irrigation, 1986-87

S. No.	*Type of Irrigation*	*Sole irrigated area (000 ha)*				
		kharif	*rabi*	*annual*	*others*	*total*
1.	Shallow tubewells	6,344	8,025	682	561	15,612
2.	Deep tubewells	942	1,496	144	182	2,764
3.	Dugwells	3,177	3,986	793	459	8,414
4.	Ground-water (1 to 3)	10,463	13,507	1,619	1,202	26,790
5.	Surface water:					
	(a) flow	3,172	1,493	131	93	4,890
	(b) lift	690	808	166	139	1,802
	Total	3,862	2,301	297	232	6.692
	All Sources	14,325	15,808	1,916	1,434	33,482

SHALLOW TUBEWELLS

Among minor irrigation sources, shallow tubewells reveal the best rate of capacity utilisation in the range of 93-107 per cent. Since the mean value of the end values of this range is 100, one may validily conclude that shallow tubewells are fully utilised, a result that is quite in harmony with the general impression of full capacity utilisation in respect of minor irrigation works as a whole. So, the legend of full capacity utilisation may be valid for this mode of irrigation only.

Among the states where *overutilisation* of shallow tubewell capacity occurred as per the CMI data was Punjab. According to the Census, the irrigation capacity of shallow tubewells in this state was for 4.660 million crop hectares. The owners of such tubewells managed to irrigate 4.490 mha solely with tubewell water, and another 1.061 mha benefited as supplementary irrigation from tubewell water--it was a case of conjunctive use of canal and tubewell waters, notwithstanding the fact that much of the groundwater pumped by tubewells was of canal origin *i.e.,* it is speed-in canal waters. Thus, the rate of capacity utilisation falls in the range of 96-119 per cent.

The emergence of fuller utilisation of capacity in the case of shallow tubewells partly arises because of under-assessment of irrigation capacity of such tubewells. The mean value of irrigation capacity per shallow tubewell in use during 1986-87, as per the CMI statistics, works out to about 3.5 hectares of crop area. This is well below the technical capability of a minimum size tubewell which can irrigate about 4 ha area in each of the two main crop seasons of an agricultural year, plus some additional 2 to 3 ha during the summer season. It appears that in the present Census the assessment of created irrigation potential is based on actual culturable command area of the owner of a tubewell, which comes to hardly 3 ha per tubewell for all the states taken.

DUGWELLS

According to the CMI, the total irrigation capacity of over 7.7 million dugwells (including 0.36 million dugwells with bore holes) was about 11.8 mha which was utilised to the extent of 71-77 per cent during 1986-87. This not very reassuring level of capacity utilisation is partly due to a substantial number of dugwells being not in use during 1986-87. In contrast to barely one per cent shallow tubewells being out of use in 1986-87, nearly one-fifth (1.5 million out of about 7.7 million) of dugwells were reportedly out of use in that year. As a result of below normal rainfall in 1986-87, some shallow wells would have possibly gone dry. This possibility could be high in Gujarat where rainfall deficiency was 50 per cent and where more than half the dugwells (0.46 million out of 0.84 million) were reportedly not in use in this state in the Census year. Another reason behind wells being out of use could be their owners' acquiring access to some newer source of irrigation which is either much cheaper (*e.g.*, canal water), or more adequate in quantity and quality (*e.g.*, a shallow tubewell of one's own). It is this factor that is behind a substantial proportion of dugwells in the Indo-Gangetic plains lying unused as their owner farmers have installed private tubewells of their own after the green revolution. The percentage of dugwells not in use was as high as 67 in the state of Punjab, 47 in Uttar Pradesh and 22 in Haryana. Finally, some dugwells might have become dysfunctional due to serious overexploitation of groundwater aquifers beneath them, including intrusion of saline waters, such as the well-known ingress of sea water into wells in the coastal region of Saurashtra in Gujarat state.

DEEP TUBEWELLS

Deep tubewells constructed in India usually extend to a depth of 100 metres and more, and are designed to give a water discharge of 100-200 cubic metres per hour (about 6 to 8 times that of a shallow tubewell). In India these are synonymous with state/public tubewells. Out of nearly 104 thousand deep tubewells nearly 8 thousand (nearly 8 per cent) were not in use. The Census picture of 72-81 per cent capacity utilisation in respect of them turns out to be far less dismal than known through earlier survey and evaluation studies on state tubewells. Therefore, this Census finding needs to be taken with due circumspection. For example, statistics for public tubewells for 1983-85 period, as reported by R.S. Saksena, the former chief of Minor Irrigation Division of the Ministry of Water Resources, reveal as low as 6 per cent utilisation in Bihar, whereas the Census finding in this regard is 80 per cent utilisation of capacity. True, the measurement of capacity utilisation in survey reports is often done on the basis of performance as per the annual running hours per tubewell. And it is known to be much less satisfactory than the one based on irrigated area statistics. Finally, the rather good capacity utilisation (nearly 72 per cent) for deep tubewells in Gujarat state is really not in respect of government tubewells. For, bulk of the 23,000 deep tubewells of this state are actually in the private sector.

The CMI data shows comparatively high rate of capacity utilisation (71 per cent) for 25,000 deep tubewells in U.P., for which other studies place the same rate in the neighbourhood of 40 per cent. This divergence is due to the peculiar practice of the Irrigation Department of this state, whereby area under long duration crops, notably sugarcane, is counted more than once. So, the agricultural year is divided into three 4-monthly segments (kharif, rabi and summer, *i.e.* 'zaid'), and area benefited from a public tubewell is taken as the sum of these season-wise areas. Thus one crop hectare under sugarcane gets reported as three hectares while compiling utilisation statistics.

SURFACE FLOW IRRIGATION

The rate of capacity utilisation for this class of works falls in the range of 77-83 per cent. As many as 50,000 flow irrigation schemes were found to be not in use during the Census year. Thus nearly 11 per cent of such schemes were not in use at all in 1986-87. This percentage was nearer 31 in the case of Karnataka state, for which we know from another investigative study by the Indian Institute of Management, Bangalore, that "over half of the estimated 43,000 odd tanks in Karnataka state were found to be more or less completely silted up, resulting in 2.5 lakh ha area being lost to irrigation (Source: *Report of the Working Group on Minor Irrigation for the Eighth Plan 1990-95, p.17)*. This means that a transient factor like rainfall deficiency could not be blamed for such silted-up tanks being out of use during 1986-87.

Most of the flow irrigation works are not individually owned the way dugwells and shallow tubewells are. According to the Census, nearly 54 per cent of these are owned by government departments, 28 per cent by farmers or their societies, and the rest (18 per cent) by public or community organisations. The last one is a predominant mode of ownership in Nagaland, while farmers and their societies are preponderant in West Bengal. Government ownership of flow irrigation works is predominant in southern states and Madhya Pradesh.

A majority of the minor flow irrigation schemes are based on storage works, and the rest are based on stream diversion--both permanent and temporary diversion works are equal in numbers. More precisely, nearly two-thirds of the 461 thousand flow irrigation works in use have storage backup. This percentage is higher in the southern states known for tank irrigation since time immemorial: 90 per cent for Andhra Pradesh, 87 per cent for Tamil Nadu and 73 per cent for Karnataka. Diversion works are predominant in Himachal Pradesh (95 per cent), Jammu and Kashmir (99 per cent), Nagaland, Sikkim and hilly parts of U.P. and of some other states.

SURFACE LIFT IRRIGATION

Lastly, we come to lift irrigation schemes based on surface water sources such as rivers, natural and man-made drains, canals, tanks, and large reservoirs. Since the LUS and other published data sources provide little information on this upcoming source of irrigation, we discuss the CMI results in this regard in greater detail.

According to the Census, there were over 493 thousand such lift· irrigation works. Most of these were found to be in use during 1986-87--those out of use constituted only about 2 per cent of the total in existence. West Bengal accounts for almost two-fifths of such schemes. States like Maharashtra and Madhya Pradesh, where modern lift irrigation schemes involving the use of power pumpsets have come up in the recent period, account for hardly one-tenth each in the national total: 66 thousand for Madhya Pradesh, 57 thousand for Maharashtra and 52 thousand in Kerala. These three states, however, account for nearly 69 per cent of all the 176 thousand lift schemes operated with the help of electric power, the cheapest source of energy for farm operations in India. Diesel-operated schemes, numbering over 210 thousand, were predominantly found in West Bengal (104 thousand), Uttar Pradesh (34 thousand) and Madhya Pradesh (34 thousand), mainly because of lower development of rural electrification in these states. Nearly 20 per cent of the lift schemes were operated with animate power derived from animals or from human beings, or both. Such traditional lifts generally entail low heights through which water is lifted for irrigation. In contrast, lifts based on power pumpsets are used for heights upto 50 metres (much more in the case of lifts meant for civic uses of water).

A majority of the lift schemes are operated by farmers and societies. Their ownershipwise breakup is as follows:

1.	private/farmers' societies	402,789
2.	government	46,975
3.	public organisations	41,281
4.	unspecified	2,931
	Total	493,976

A majority of the lift schemes are based on water stored in tanks, followed by flowing water in river and in drains and canals as shows below:

1.	lift schemes based on tanks	242,238
2.	lift schemes based on rivers	191,331
3.	lift schemes based on drains and canals	50,150
	Total (1 to 3)	483,719

The aggregative picture shown above turns around once we exclude West Bengal state from totality. That is to say, for all-India excluding West Bengal rivers constitute the major source of water for lift schemes. This is clear from the following data:

	No. of lift schemes by source of water supply (000)			
	river	*drain/canal*	*tank*	*total*
1. West Bengal	12	22	168	203
2. Rest of India	179	28	78	281

Though the phenomenon of lifts being out of use is quite low at the all-India level, it is of considerable significance in some states. For example, one-fifth of about 1500 lifts of Gujarat were not in use during 1986-87, presumably because of low availability of water in rivers, tanks, drains etc. due to severe drought in this year. In Orissa, one-eighth of the lifts were reportedly not in use during the reference year. To what extent this was due to pumpsets being out of order during 1986-87 is not known (both in Orissa and Gujarat lift schemes are wholly pumpset operated). Anyway, the rates of capacity utilisation in these two states were much lower than the aggregate rate of 74-86 per cent (the rate was 27 per cent in the case of Gujarat lifts and 41 per cent for the Orissa lifts). The low utilisation in the case of Gujarat might be mainly due to the drought factor. In the case of Orissa, a NABARD study *(Public Tubewells and River Lifts in Orissa*, 1984) also revealed poor rate of capacity utilisation during the following three years prior to the reference year of the study.

	% utilisation of capacity	
Year	*Kharif*	*Rabi*
1978-79	31	31
1979-80 (drought year*)	45	57
1980-81	36	38

* 54% rainfall deficiency.

Maharashtra's data reveals an exceedingly good performance of lift capacity utilisation. The created irrigation potential for about

57,000 lift schemes, all fitted with pumpsets (81 per cent electric ones), is of the order of 296 thousand hectares. In sole irrigation, these lifts benefited total crop area to the extent of 286 thousand ha. Also, these provided supplementary irrigation to another 133 thousand crop hectares. So, in all 419 thousand ha were irrigated during 1986-87, giving rise to a situation of overutilisation of capacity. Such overutilisation, questionable as it is, emerges despite heavily irrigated annual crops like sugarcane claiming an important share in total benefited area, as is revealed by the following data:

(000 ha)

	Kharif area	*Rabi area*	*Annual crops area*	*Others*	*Total area*
1. Sole irrigation	83	96	92	16	286
2. Supplementary irrigation	23	103	6	2	133
Total lift irrigation	106	199	98	18	419

SUMMARY AND CONCLUSIONS

Minor irrigation has acquired a halo of sorts about itself even in intelligent circles. It is time that admirable qualities attributed to small-scale irrigation works are duly cross checked with ground realities. In a small way, the Report of the Census of Minor Irrigation Schemes 1986-87, released by the Ministry of Water Resources (MOW), Government of India, affords us such an opportunity. While it is not possible to ascertain the validity of the widely-held view that minor irrigation is a low cost option for the Indian economy, we can assess with the help of this census the ground truth with regard to the legend that minor irrigation capacity ('potential' in the planners' parlance), unlike capacity of major and medium irrigation (*i.e.,* large-scale irrigation), is fully utilised.

Admittedly, the objective of the above-mentioned Census was much more than merely testing some of the assumptions of the planners with regard to minor irrigation development in India. The Census does add to our data base on minor irrigation in important ways. As for example, we get from this Census a useful district level picture of minor irrigation--such information can be partially obtained from state-level statistical documents, but not at all from the plan documents. Again, the Census provides, for the first time, information about (i) conjunctive use of irrigation sources in India, (ii) the extent of irrigation structures having fallen in disuse over time and (iii) seasonal orientation of minor irrigation—it is an important parameter that determines the production impact of an irrigation source in the Indian monsoonal conditions of farming.

The idea of conducting a nationwide census of minor irrigation works was first mooted by the Indian Planning Commission in 1970 as the Commission found that, in the absence of a nodal agency for maintaining statistics relating to minor irrigation development, hardly any State could provide a reliable picture about the state of minor irrigation segment in its entirety in the State as a whole. The census was finally taken up as part of the Scheme for Improvement of Irrigation Statistics which was launched in 1980-81. With the singular exception of Rajasthan, all States and the Union Territories carried out this census, with 1986-87 as the reference period (1987-88 in the case of West Bengal). Each State was expected to publish the census results about its own state. Our analysis, however, is based on the nationwide data released by the MOW Report mentioned earlier. It may be pointed out that in this Report there is neither any accompanying text nor analysis expounding the numerous tables contained in this Report.

The CMI Report is rich in information. We have selected a few aspects for observation and comment here in this chapter. Our major findings are as follows.

Leaving aside the complex issue of measurement of irrigation capacity, one is struck by the fact that the Census estimate or figure in respect of total minor irrigation capacity for 1986-87 turns out to be materially no different from that prepared by the Planning Commission on the basis of discussions with the State representatives at the time of annual plan parleys. Unexpectedly, the Plan estimate turns out about 8 per cent below the Census figure of about 42 mha. This would suggest that Indian planners have not been overprojecting ground reality in respect of achievements in minor irrigation capacity (we are making a tacit assumption that the Census data reflects ground truth).

But the this assertion loses much of its force the moment we do a disaggregative analysis by state and by category of minor irrigation. For almost half the States the difference between the two sources of data is in one direction, and for the rest in the other direction. Two striking examples of such differences are for the two States of Gujarat and Uttar Pradesh. In U.P., the Census figure of minor irrigation capacity is around 10 mha, about 3.8 mha less than the Plan figure of 13.847. In contrast, for Gujarat the Census figure of created minor irrigation potential exceeds the corresponding Plan estimate by a huge margin (the Census figure is 3.844 mha compared to 1.728 mha in the Plan statistics). We are, therefore, forced to conclude that both the Census and the Plan figures in respect of minor irrigation potential lack realism as the very concept of irrigation potential is not well defined or definable in an operational sense. Hence, a priori one cannot affirm that the Census estimate is necessarily a more reliable measure of minor irrigation potential. In short, a census of minor irrigation, even if conducted every five years, cannot serve as a useful cross-check on planners' own, normative assessment of minor irrigation potential.

Notwithstanding the intractable problem of measuring satisfactorily capacity of a small-scale irrigation work, one finds that the Indian minor irrigation segment is not free from the incidence of underutilisation of capacity. At the aggregate national level, somewhere between 9 and 17 per cent of the minor irrigation potential established by 1986-87 was unutilised in that year. The lower bound of this range estimate is based on the assumption of counting fully all the supplementary area benefited by such works when these are used in conjunction with either major and medium irrigation works or among themselves, that is, say, a farmer uses irrigation from both a dugwell and a tank in watering a given plot of his land. The upper bound of 17 per cent underutilisation of capacity excludes such supplementary area and is based on sole irrigated area statistics only. Even this level of underutilisation is almost twice the level conveyed by statistics in our Plan documents.

Among the five principal categories of minor irrigation works, the incidence of underutilisation of capacity is mainly found in the case of dugwells, deep tubewells, surface flow and lift irrigation schemes. It is in the case of shallow tubewells that one discerns signs of full utilisation of capacity. Thus, the widespread impression that minor irrigation capacity, unlike major and medium irrigation capacity, is fully utilised is valid for private shallow tubewells only, not for other categories of minor irrigation works.

The extent of capacity lying idle among four types of minor irrigation works varies as follows:

1. dugwells	23-29 per cent
2. surface flow irrigation	17-23 per cent
3. surface lift irrigation	14-26 per cent
4. deep tubewells	19-28 per cent

The estimated rate of capacity utilisation for deep tubewells appears to be seriously biased upwards and should be taken with circumspection, more so if it is to be used as a measure of performance of government owned or public tubewells of the country. Below par utilisation in their case can be attributed to (i) poor management under public aegis and (ii) inadequacies of electricity supply system. The factor (i) applies also to minor surface irrigation works, especially tanks, maintained largely by government or community agencies like panchayats. Another factor behind underutilisation is the sizable number of works being not in use during 1986-87, partly because of drought conditions. The severe drought in Gujarat put almost half the minor works out of use during 1986-87. This incidence of works not found in use was higher in the case of dugwells, one-fifth of them being not in use during 1986-87. While shallow dugwells could be out of use because of drought rendering them as mere dry holes, many could have fallen into disuse because of their owners acquiring access to other superior irrigation sources. In Punjab, U.P. and Haryana the high incidence of dugwells not in use during 1986-87 could be linked to the pervasive rise of private tubewell irrigation in these states. By category of works the incidence of disuse varied as follows:

	Per cent out of use
1. dugwells	19
2. surface flow works	11
3. deep tubewells	8
4. shallow tubewells	2
5. surface lift irrigation	1
6. all works	13

Electric pumpsets are now the main motive force in the operation of minor irrigation structures. According to the Census, about 4.9 million electric pumpsets were installed on minor irrigation structures, and nearly 94 per cent were fitted to wells of various types. It appears this number is markedly less than the official number of electric pumpsets reported in Plan documents, namely, 6.4 millions. The more noteworthy result from the Census is about acute power shortage faced by pumpset owners who hardly were supplied with 4 hours of electricity per day (of 24 hours) during the peak time of crop seasons of 1986-87. It is pointless to expand the number of minor irrigation structures without commensurate expansion in electricity supply system (power generation and rural electrification). Otherwise, existing power would either be spread thinly among the pumpset owners, or more and more of it would be diverted to the farm sector—thereby imposing heavy opportunity costs on other sectors of the economy.

APPENDIX NOTE

METHODOLOGY BEHIND MINOR IRRIGATION CENSUS

The Scheme for Improvement of Irrigation Statistics was launched in 1980-81 by the Department of Agriculture, Government of India. One of the objectives of the scheme was to conduct census of minor irrigation works on a quinquennial basis. Suitable number of staff were also proposed to be created in each State for compilation of minor irrigation statistics. Upto 1985-86, only 15 States implemented this scheme and appointed staff for the purpose.

At the beginning of the Seventh Five Year Plan, Planning Commission insisted on preparing a sound data base for future planning of minor irrigation sector. Immediate action was initiated by the Ministry of Water Resources to conduct a census of all minor irrigation works. To avoid any delay, the entire funds for the purpose were provided by the Centre to the States and Union Territory administrations. A meeting of the officers incharge of minor irrigation programme in the major States was convened to discuss the various issues relating to census operation. The meeting discussed in detail the machinery to be involved, requirement of funds and information to be collected during the census. The details of the operation of Agricultural Census 1980-81 were also considered. Based on the norms and the mechanism devised after a number of meetings with the concerned Central Government Departments, the scheme of first census was finally formulated with the following objectives:-

(a) to enumerate completely the sources of minor irrigation in the States/Union Territories.

(b) to assess sources-wise area irrigated during Kharif, Rabi and summer seasons.

(c) to assess the contribution of these minor irrigation works by way of new irrigation or as supplementary irrigation source.

The census scheme was finally sanctioned on 13th March, 1987 for a total estimated cost of Rs. 3.5 crores. The reference year of the census was decided to be 1986-87. Each State and UT administration was requested to identify a nodal Department for conducting the census in its respective State/UT. The proposals for the nodal Deptt. of the States/UTs administration indicating the estimated cost in a specified format were received and scrutinised by the Minor Irrigation Division of the Ministry and placed before the sanctioning Committee for sanction of the amount required. 50 per cent of the estimated cost as sanctioned was released to the States/UTs in advance immediately after receiving the proposal. 25 per cent was released as soon as the primary enumeration work was completed and the balance 25 per cent after receiving the districtwise census abstract in the Minor Irrigation Division of the Ministry.

A small Advisory Committee under the chairmanship of Additional Secretary (MOWR) was constituted to oversee and monitor the implementation of the Census. The representatives of the Planning Commission, Central Statistical Organisation, Department of Agriculture, and Department of Rural Development were the other members of the Committee.

CONCEPTS AND DEFINITIONS

Minor Irrigation Schemes were classified into 5 broad categories:-

1. **Dugwells** (including dug-cum-bore wells).
2. **Private shallow tubewells.** In sedimentary formations depth of a shallow tubewell does not exceed 60-70 metres. These tubewells are either cavity tubewells or strainer tubewells. The tubewells give yield of 100-300 cubic metres per day which is roughly 2 to 3 times that from a dugwell.
3. **Deep tubewells.** These usually extend to a depth of 100 metres and more and are designed to give a discharge of 100 to 200 cubic metre per hour. Their annual output of water is roughly 15 times of an average shallow tubewell.
4. **Surface flow irrigation schemes.** These schemes use the rain water for irrigation purposes either by storing it or by diverting it from a stream, nala or river. Sometimes, permanent

diversions are constructed for utilising the flowing water of a stream or river. These schemes are of three kinds:

(i) Storage schemes

Storage schemes include tanks and reservoirs which impound water of streams and rivers for irrigation purposes. The essential features of these schemes are (1) a bund or a dam which is generally earthen but also sometimes partly or fully masonary, (b) anicuts and feeder channels to divert water from adjoining catchments, (c) a waste weir to dispose of surplus flood water, (d) sluice or sluices to let out water for irrigation, and (e) conveyance and distribution system. The size of the storage is determined by the run-off expected on the basis of dependable monsoon rainfall in the catchment and by the fact whether the rainfall and cropping pattern would permit more than one filling of the tank.

(ii) Diversion schemes including hilly channels

The schemes aim at providing gravity flow irrigation by mere diversion of stream water supply without arranging any storage. Essentially, such a scheme consists of two parts: (a) an obstruction (weir) or bund constructed across the stream for raising and diverting water. The weir is called anicut in the South, bandhara in Maharashtra and Gujarat, and thingal in the Assam region, and (b) an artificial channel known as kuhl in the hilly areas, pyne in Chhota Nagpur and Bihar, dongs and ilhowkongs in the Assam region and so on.

(iii) Water conservation-cum-groundwater recharging schemes

Under this head are included schemes which serve primarily one or more of the following purposes: (a) submerging agricultural land during monsoon for sowing post-monsoon crops, (b) improving moisture regime of the adjoining fields downstream for raising of post-monsoon crops without irrigation, and (c) replenishing the groundwater.

5. **Surface Lift Irrigation Schemes.** In regions, where the topography does not permit direct flow irrigation from rivers and streams, water has to be lifted into the irrigation channels.

These works are similar to those of diversion schemes, but in addition pumps are installed and pump houses constructed.

The statistics of irrigation mainly include data on area irrigated by different sources and under different crop seasons. Select definitions which are prevalent in respect of irrigated areas are as follows:

Gross command area: The total gross area which can be physically irrigated from a scheme without considering the limitations of the quantity of water available.

Culturable command area: It is the gross area, less the area of unculturable land including habitation area, ponds and land otherwise unfit for cultivation of the gross area.

Gross irrigated area: The total irrigated area under various crops during a year, counting the area irrigated under more than one crop during the same year as many times as the number of crops grown.

Gross irrigation potential created by a scheme: The total gross area proposed to be irrigated under different crops during a year by the scheme. The area proposed to be irrigated under more than one crop during the same year is counted as many times as the number of crops grown.

Irrigation potential utilised during a year: The gross irrigation potential created by the scheme as actually used during a particular year.

Supplementary irrigation: The gross cropped area is actually irrigated by two or more different sources of irrigation. All these different sources of irrigation are providing supplementary irrigation to the area.

The Census of minor irrigation works is to include all minor irrigation schemes situated in rural areas, urban areas and forest areas. The schemes which are used for drinking water purposes in addition to irrigation are to be covered under census. The minor irrigation schemes in the commands of major and medium irrigation schemes were also to be enumerated and come under the purview of the Census.

PROCEDURE FOR CONDUCTING CENSUS OPERATION

The primary enumerators while filling in the schedules were to visit the owner of the minor irrigation scheme or its next neighbour and collect information on the basis of personal enquiry from him. The physical verifications of the schemes were also to be taken up by the enumerators. Certain information relating to the schemes are to be collected by the enumerators by physical examination of the scheme.

Training-cum-workshops were to be organised by the Minor Irrigation Census Commissioner at the State headquarters in which the district level concerned officers were to participate. The details of the methodology to be adopted for the Census, its procedure, concepts and definitions etc. were to be discussed thoroughly and necessary clarifications given. The instructions for filling up the primary enumeration schedules alongwith the format for data consolidation at various stages were also to be discussed. In turn, the district level officers were to organise training-cum-workshops at the district headquarters where the primary schedules and the procedures for it were to be explained to enumerators.

MINOR IRRIGATION CENSUS OPERATION

A brief note giving the broad features of the scheme was circulated to the States/UTs in February, 1987. Stress was given on identifying a nodal department for undertaking the work with a view to completing it by 1987-88. The pre-testing of schedules were conducted in the States of Andhra Pradesh, Haryana, Himachal Pradesh, Maharashtra, Uttar Pradesh and West Bengal. In consultation with the Advisory committee the primary enumeration schedules and the formats for consolidation of data were prepared and circulated to the States in August, 1987. The following time schedule for census operation was intimated to the States in June, 1987.

The States/UTs were requested to submit their proposals alongwith the cost estimates in the prescribed format. During 1986-87 the proposals were received only from the State Governments of Andhra Pradesh, Haryana, Himachal Pradesh, Kerala, Sikkim, Tamil Nadu, West Bengal and Uttar Pradesh. These proposals were scrutinised and the administrative approval for conducting the census at the estimated cost were issued alongwith 50% Central assistance released in advance. Rest of the State Govts. and UTs Administrations submitted their proposals in subsequent years.

Sl.No.	*Operation*	*Date by which to be completed*
1.	Pre-testing of schedules in selected states	14th August, 1987
2.	Communication of final schedules to the States/UTs	31st August, 1987
3.	Printing to format and training to personnel.	15th November, 1987
4.	Primary enumeration	15th January, 1988
5.	Compilation of final figures at block level after completing the sample check	15th January, 1988
6.	Compilation of final figures at district level	31st January, 1988
7.	Compilation of final figures at State level and forwarding these to the Ministry of Water Resources	28th February, 1988

The districtwise abstracts of the census were to be supplied to the Minor Irrigation Division of the Ministry of Water Resources, Government of India. The States were also requested to bring out a minor irrigation census report in respect of their State, giving blockwise data relating to minor irrigation works. The tabulation plan for such a report was also sent to the Minor Irrigation Census Commissioners.

The National Informatics Centre (NIC), Department of Science and Technology, provided computerisation facilities at the district headquarters. A 2-day workshop was organised by the Ministry of Water Resources in association with the National Informatics Centre, Delhi in May, 1989. The Minor Irrigation Census Commissioners of the States participated in the workshop.

LIMITATIONS

The Census of Minor Irrigation Works has been conducted for the first time in the country and completed by the states on various dates in a span of about four years. A number of difficulties were encountered by the states. Depending on the gap between the reference year and the date of census the reliability of data varies. Smaller the gap, more reliable the data collected. Despite best efforts by the minor irrigation census commissioners in the states, certain limitations remain.

1. Census of minor irrigation works could not be completed in Rajasthan. The work has been held up due to non-cooperation of patwaries, the primary enumerators.
2. West Bengal has conducted the Census of Minor Irrigation Works with reference to 1987-88 whereas other states have conducted it with reference to 1986-87.
3. Year-wise installation of minor irrigation works has not been collected during the census by some of the states such as U.P., Punjab and West Bengal.
4. The enumeration of groundwater structures in Kerala has not been taken up. Only the estimated number of dugwells in the state which are also used for irrigation has been estimated based on sample surveys. The details in respect of these dugwells are not available.
5. The State Government of Karnataka has not enumerated minor irrigation works which are installed in the commands of major and medium irrigation projects.
6. Distribution of dugwells and private shallow tubewells among the farmers of different holding size groups and the farmers of different social status has not been collected for Andhra Pradesh.
7. Distribution of groundwater structures by the farmers in different holding size groups and social status etc. is not available for all structures, specially in the case of Maharashtra.
8. The culturable command area (CCA) of a minor irrigation structure has been higher than the gross irrigation potential created by it. Normally it should have been equal or less. Even if single irrigation crop is proposed to be grown during a year in the culturable command area of the scheme the gross irrigation potential created by it is equal to the culturable command area of the scheme. If more than one irrigated crop is proposed to be grown during a year the gross irrigation

potential created would be higher than the CCA. Some of the states have intimated that due to less storage or less discharge the irrigation potential has been less but this may happen in a particular year and affect the utilisation of gross irrigation potential created by the scheme in that year. In case of a reduction in the capacity of the scheme, such as due to siltation in case of a tank, there is a shrinkage in the command itself and accordingly of the other parameters.

9. The actual utilisation of gross irrigation potential of a scheme in 1986-87 has been higher than the gross irrigation potential created by it. If the full capacity is being utilised in a particular year, the gross irrigation potential created is equal to the actual area irrigated by the scheme. In case of a reduction in the storage or less discharge, the actual area irrigated during that year would be less than gross irrigation potential created by the scheme.

10. The minor irrigation structures which are providing supplementary irrigation during 1986-87 have been showing higher area irrigated by other sources in the commands of the structures. Actually, it would have been at the most equal to the gross area irrigated by the scheme during the year. The scheme has provided supplementary irrigation only. Generally, a part of the gross area actually irrigated by a scheme is also being irrigated by some other source.

INDEX

D

H

I

N

O

P

Y

Z

Institute of Economic Growth, Delhi
Studies in Economic Development and Planning

1. *The Role of Small Enterprises in Indian Economic Development* by P.N. Dhar and H.F. Lydall, 1961. (Asia Publishing House, Bombay).
2. *Indian Tax Structure and Economic Development* by G.S. Sahota, 1961. (Asia Publishing House, Bombay).
3. *Agricultural Labour in India* by V.K.R.V. Rao, 1962. (Asia Publishing House, Bombay).
4. *Foreign Aid and India's Economic Development* by V.K.R.V. Rao and Dharm Narain, 1963. (Asia Publishing House, Bombay).
5. *Agricultural Production Functions, Costs and Returns in India* by C.H. Hanumantha Rao, 1965, (Asia Publishing House, Bombay).
6. *Resource Allocation in the Cotton Textile Industry* by Dharma Kumar, S.P. Nag and L.S. Venkataramanan, 1965. (Asia Publishing House, Bombay).
7. *India's Industrialisation and Mineral Exports* by V.P. Chopra, 1965. (Asia Publishing House, Bombay).
8. *Taxation of Agricultural Land in Andhra Pradesh* by C.H. Hanumantha Rao, 1966. (Asia Publishing House, Bombay).
9. *Some Aspects of Cooperative Farming in India* by S.K. Goyal, 1966. (Asia Publishing House, Bombay).
10. *Demand for Energy in North-West India* by P.N. Dhar and D.U. Sastry, 1967. (Asia Publishing House, Bombay).
11. *Some Aspects of the Structure of Indian Agricultural Economy 1947-48 to 1961-62* by P.V. John, 1968. (Asia Publishing House, Bombay).

12. *Wages and Productivity in Selected Indian Industries* by J.N. Sinha and P.K. Sawhney, 1970 (Vikas Publishing House, Delhi).

13. *Inventories in Indian Manufacturing* by K. Krishnamurty and D.U. Sastry, 1970. (Academic Books Ltd., New Delhi).

14. *Buffer Stocks and Storage of Major Foodgrains in India* by A.M. Khusro, 1973. (Tata-McGraw Hill Publishing Company Ltd., Bombay).

15. *Bangla Desh Economy: Problems and Prospects* edited by V.K.R.V. Rao, 1972. (Vikas Publishing House, New Delhi).

16. *The Economics of Land Reform and Farm Size in India* by A.M. Khusro, 1973. (Macmillan Co. of India Ltd., Madras).

17. *Technological Change and Distribution of Gains in Indian Agriculture* by C.H. Hanumantha Rao, 1975. (Macmillan Co. of India Ltd., Madras).

18. *Investment and Financing in the Corporate Sector in India* by K. Krishnamurty and D.U. Sastry, 1975. (Tata McGraw Hill Publishing Company Ltd., New Delhi).

19. *Land Reforms in India: Trends and Perspective* by P.C. Joshi, 1975. (Allied Publishers Private Limited, Bombay).

20. *Cost Benefit Analysis: A Case Study of the Ratnagiri Fisheries Project* by S.N. Mishra and John Beyer, 1976. (Hindustan Publishing Corporation, Delhi).

21. *Livestock Planning in India* by S.N. Mishra, 1978 (Vikas Publishing House, Delhi).

22. *Rice Marketing System and Compulsory Levies in Andhra Pradesh* by K. Subbarao, 1978. (Allied Publishers Private Limited, Bombay).

23. *Farm Size, Resource - Use Efficiency and Income Distribution* by G.R. Saini, 1979. (Allied Publishers Private Limited, New Delhi).

24. *Class Relations and Technical Change in Indian Agriculture* by Sipra Dasgupta, 1980. (Macmillan Co. of India Limited, Delhi).

25. *Conglomerates and Multinationals in India: A Study of Investment and Profit* by N.S. Siddharthan, 1981. (Allied Publishers Private Limited, New Delhi).

26. *Development of Tubewell Irrigation in India* by B.D. Dhawan, 1982. (Agricole Publishing Academy, New Delhi).

27. *Economic Appraisal of Irrigation Projects in India* by Basawan Sinha and Ramesh Bhatia, 1982. (Agricole Publishing Academy, New Delhi).

28. *Economic Planning for Iron Ore in India* by Gopal Kadekodi, 1982. (Hindustan Publishing Corporation, Delhi).

29. *Oilseeds Economy of India* by Uma Kapila, 1982. (Agricole Publishing Academy, New Delhi).

30. *Dry Farming Technology in India : A Study of its Profitability in Selected Areas* by P. Rangaswamy, 1982. (Agricole Publishing Academy, New Delhi).

31. *Planning for the Petroleum and Fertilizer Industries: A Programming Model for India* by Ramesh Bhatia, 1983. (Oxford University Press, Delhi).

32. *Studies in Social Dynamics of Primary Health Care* by Ashish Bose and P.B. Desai, 1983. (Hindustan Publishing Corporation, Delhi).

33. *Rural Development Planning : Design and Method* by S.N. Mishra, 1984. (Satvahan Publications, New Delhi).

34. *Cotton Mill Industry in India* by D.U. Sastry, 1984. (Oxford University Press, Delhi).

35. *Economics of Agricultural Credit* by Arun Kumar Bandyo–padhyay, 1984. (Agricole Publishing Academy, New Delhi).

36. *Slowdown Global Economic Maladies* by A. Brody, 1985. (Sage Publications India Private Limited, New Delhi).

37. *Public Expenditure, Inflation and Growth : A Macro-Econometric Analysis for India* by B.B. Bhattacharya, 1984, (Oxford University Press, Delhi).

38. *Intensification of Agriculture : A Study in the Plains of Uttar Pradesh* by S.K. Ray, 1985 (Hindustan Publishing Corporation, Delhi).

39. *Productivity Growth in Indian Industry* by B.N. Goldar, 1986 (Allied Publishers (P) Ltd., New Delhi).

40. *Cold Hearths and Barren Slopes : The Woodfuel Crisis in the Third World* by Bina Agarwal, 1986. (Allied Publishers (P) Ltd., New Delhi).

41. *Rural to Urban Migration and the Urban Labour Market* by Biswajit Banerjee, 1986. (Himalaya Publishing House, Delhi).

42. *Economics of Groundwater Irrigation in Hard Rock Regions—With Special Reference to Maharashtra State* by B.D. Dhawan, 1986. (Agricole Publishing Academy, New Delhi).

43. *Agricultural Price Policy and Income Distribution in India* by Alain De Janvry and K. Subbarao, 1986. (Oxford University Press, New Delhi).

44. *The Economics of Inequality in Education* by J.B.G. Tilak, 1987. (Navrang Publishers & Book Sellers, New Delhi).

45. *Nutritional Planning in India* by Rajaram Dasgupta, 1989. (Navrang Publishers & Book Sellers, New Delhi).

46. *Irrigation in India's Agricultural Development* by B.D. Dhawan, 1988 (Sage Publications, New Delhi; Second Edition by Commonwealth Publishers, New Delhi, 1994).

47. *Unstable Agriculture and Droughts : Implications for Policy* by C.H. Hanumantha Rao, Sushanta K. Ray and K. Subbarao, 1988. (Vikas Publishing House, New Delhi).

48. *Fertility in India* by S.C. Gulati, 1988 (Sage Publications, New Delhi).

49. *Planning for Coal Sector* by G. Kadekodi, 1988. (B.R. Publishing Corporation, Delhi).

50. *Marriage Age in India* by R.P. Goyal, 1988 (B.R. Publishing Corporation, Delhi).

51. *Redistributive Taxation and Public Sector Pricing* by M.N. Murty, 1989 (Commonwealth Publishers, New Delhi).

52. *Participatory Development : People and Common Property Resources* by Kanchan Chopra, Gopal Kadekodi, M.N. Murty, 1989. (Sage Publications, New Delhi).

53. *Studies in Irrigation and Water Management* by B.D. Dhawan, 1989. (Commonwealth Publishers, New Delhi).

54. *Livestock Development in India : An Appraisal* by S.N. Mishra & Rishi K. Sharma, 1990 (Vikas Publishing House (P) Ltd., New Delhi).

55. *Agricultural Development in Punjab : Issues in Resource Use and Sustainability* by Kanchan Chopra, 1990. (Vikas Publishing House (P) Ltd., New Delhi).

56. *Studies in Minor Irrigation: With Special Reference to Groundwater* by B.D. Dhawan, 1990 (Commonwealth Publishers, New Delhi).

57. *Interlinked Agrarian Markets in Rural India* by Kailas Sarap, 1991. (Sage Publications, New Delhi).

58. *Trends and New Tendencies in Indian Irrigated Agriculture* by B.D. Dhawan, 1993 (Commonwealth Publishers, New Delhi).

59. *Agricultural Growth, Rural Poverty and Environmental Degradation in India* by C.H. Hanumantha Rao, 1994 (Oxford University Press, New Delhi).

60. *Groundwater Depletion, Land Degradation and Irrigated Agriculture in India* by B.D. Dhawan, 1995 (Commonwealth Publishers, New Delhi).

61. *Financial Liberalisation, Banking System and Stock Markets in India* by R.N. Agarwal, 1995. (B.R. Publishing Corporation, Delhi).

62. *Himalayan Migration: A Study of the Hill Region of Uttar Pradesh* by R.S. Bora, 1995. (Sage Publications, New Delhi).

63. *Water Institutions in India: Economics, Law and Policy* by R. Maria Saleth, 1996. (Commonwealth Publishers, New Delhi.

64. *Social Reform, Sexuality and the State* ed. by Patricia Uberoi, 1996. (Sage Publications, New Delhi).